The Service Technician's
INSPECTION AND IDENTIFICATION MANUAL

A Practical Guide for
Pest Control Professionals

William H Robinson, Ph.D.
Technical Director, B&G Equipment Co.

EDITOR
Lisa Jo Lupo

GRAPHIC DESIGNER
Jamie Winebrenner

Publisher: GIE Media, Inc.
© Copyright 2014

Address all correspondence to GIE Media, Inc. at www.giemedia.com or call 800/456-0707.

Library of Congress Control Number: 2013948938

www.pctonline.com

PREFACE

Inspect what you expect...

How many times have you treated or identified something without closely looking at it or inspecting it? You were certain what it was at the time, but later found that it wasn't what you thought. By then it was too late, time and materials had been spent and perhaps wasted, and the pest or the problem remained. Worse, your competitor got it right and got the job. We all have done that at least once—made a quick decision without looking, reading, or inspecting closely. To cite noted statistician and business consultant W. Edward Deming, a good rule to follow to prevent this is: *Inspect what you expect.* Don't assume you know what it is, check to make certain.

The purpose of this manual is to make inspection and identification easier for service technicians. The first part (Chapters 1 and 2) provides information to help the technician anticipate the common pests or damage that will be found during site inspections of various sites, indoors and outdoors, and includes a small figure. This can help the technician "start" the process of looking to identify pests and eliminate non-pests found during the process. The second part (Chapters 3 through 14) is linked to the first section by the "See page..." that follows every figure. This part contains detailed information on and illustrations of more than 200 pests and damage, as well as pages on which to add new pests you may find during service. The goal is to provide a single book with guidelines for inspection and a means (illustrations, distribution maps, identification marks, and frass) of confirming what you have found.

William H Robinson
Technical Director, B&G Equipment Company
Jackson, Georgia

TABLE OF CONTENTS

TABLE OF CONTENTS

CHAPTER 1

INTRODUCTION

If I called your company, would you send out someone who would positively identify the bug before spraying to ensure success? The guy from the last company just shrugged his shoulders, said, "I dunno," and then sprayed anyway. It didn't kill the bugs, so I got my money back. But I didn't want my money back, I wanted to get rid of the bugs!

Inspection and pest identification are critical steps in a pest prevention and control program. Inspections require knowing where to look and what to look for, and they improve only with success and experience. Identifying what is found also is important, to the customer and to the treatment method. A thorough inspection of suspected harborages and feeding sites will show the extent of the infestation and where to treat. This improves application efficacy and reduces costs. Residential and commercial customers rely on service technicians to inspect before they treat and to monitor after they treat.

Hand lens. Most household and structural pests and the damage they cause can be identified without the use of magnification, because most are large and have a shape or color pattern that distinguishes them. However, some are small, and accurate identification is necessary to locate the infested harborages and/or use the best control methods. For these, a small hand lens provides enough magnification.

Internet images. A search of the Internet using the common name or the scientific name (see Species Index) will provide color images of the pests presented in this manual. These can be useful in identification when they are correct, but they are not always. When a search results in many photographs or illustrations, there often are some that are incorrect, and the biological information and control can be outdated or simply wrong. Using the scientific name for a search usually provides more technical and accurate information.

Traps. Light traps, sticky traps, and pheromone traps provide a means of continuous inspection, as well as information on pest presence and abundance. These can provide the same level of information as would a visual inspection if the trap is carefully examined. However, glueboards and sticky traps often receive a quick look, and then are discarded or replaced. But even when the traps capture only a few individuals, they can give an early warning of a larger problem. Phorids on a light-trap glueboard may indicate a sewer problem; flesh flies may mean the air curtain is not working properly; and Indian meal moths may indicate infested food material. Traps provide information on abundance, location, and species for the technician who takes the time to look at what has been captured.

Page format. The pages in the second half of this manual are arranged to provide quick identification of the pest, along with a brief summary of its biology and habits. In some cases, there is information on the known distribution of the pest. Common and scientific names are listed in the Species Index at the back of the manual. When searching the Internet for information on the pests presented here, it is best to consider

Hand lens

Sticky trap

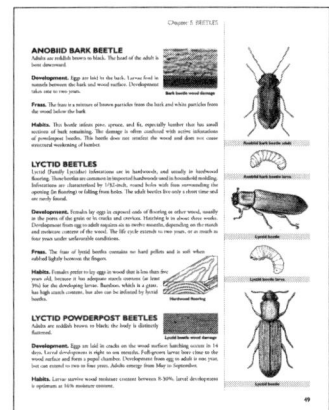

Page format

searching both names. There may be more information available when the scientific name is used, while searching on the common name may produce more images. It also is important to note that the distribution maps for some pests in this manual are provided as guidelines. The actual distribution of any insect, spider, or vertebrate is not defined by state borders, but by suitable habitats.

Other pests. Each chapter includes additional blank spaces in which pests can be added that were not included in this manual. There may be insect or vertebrate pests that are locally important, or an introduced species may spread to new regions. When this happens, there usually is an article in *Pest Control Technology* or a newsletter that includes a review of the biology and habits, and a photograph. The blank pages provide the opportunity to add this new pest information to the manual, so it can be easily found in the future.

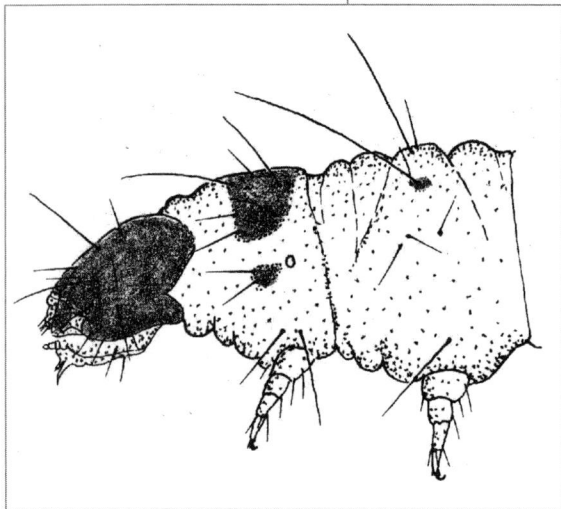

Indian meal moth caterpillar

Life stages. For many of the pests in the manual, an illustration of a larval or caterpillar stage is provided. In the case of carpet beetles and some flour and fabric pests, the larval stage may be found more often than the adult stage. The larvae of carpet beetles are distinct and can be used to identify some species. The caterpillars for flour moths and clothes moths have distinct features, but require magnification to identify. Although identification down to the species usually is not necessary, these illustrations can be useful when species matters. However they are not always helpful; for example, the maggot stage for common fly species is not easily used for identification. And while distinguishing phorid and fruit flies can be important, the pupal cases can be used for that. For wood-infesting insects, especially beetles, the adult emergence hole is used for identification, since the larval stages are in the wood and the adult stages are small, short lived, and rarely found.

Illustrations. Many illustrations in this manual were drawn by Gene Wood (1932-2013) and include his signature. These were selected from a collection of his illustrations over many years. Wood was an urban entomologist and professor emeritus at the University of Maryland. During his career at the university, he conducted research on cockroaches, termites, and integrated pest management programs. Most important was his work with professional pest control operators; he founded the Interstate Pest Control Conference. Trained as a taxonomist, Wood worked as an entomologist, but, in his heart, he was an artist and illustrator of insects. He helped a lot of us with technique and led the way with the quality of his work.

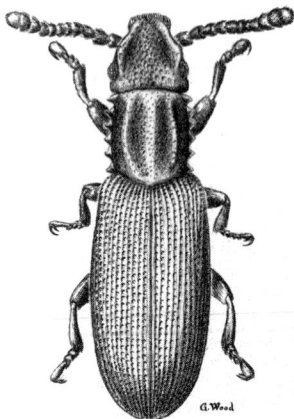

Sawtoothed grain beetle

CHAPTER 2

INSPECTION SITES

The first step in a pest management program is an inspection of the site to determine the location and population of the target pest. With this information, chemical and non-chemical methods can be used to get control. Without a thorough inspection, control may be limited or short-lived, and time and material wasted. But the initial inspection can provide information on conditions in the indoor or outdoor habitat that are favorable to the existing pests or others that may move in later. Prevention usually is based on making non-chemical changes to the habitat, and these are based on knowledge of the food and harborage requirements of insect and rodent pests.

EVIDENCE

The procedure for residential and commercial inspections is to search for the presence or other evidence of the target pest. The time devoted to inspections is often limited, and it may not overlap pest activity periods or the life stages that are the most active. For example, rats, mice, and cockroaches are active at night; Indian meal moth caterpillars may not be detected until they leave the infested material to find a place to pupate. In many cases it is only the evidence, such as the cast skins of carpet beetles or feces of rodents that remain for identification. It is the trained and experienced technician that can distinguish signs of pests and where to look for them.

Expected vs. Found. The insects, spiders, and vertebrates that may be found during any inspection depend on the time, day, season, and conditions at the location. Pests seen by a customer one day may not be seen on the day the service technician arrives to conduct an inspection. But there may be signs or other evidence of the pest, even if it is not seen. The pests listed for the locations presented in this manual represent what is possible or what could be expected. It is unlikely that all those listed would be found at any one time at one location, but this provides a range of possibilities for which service technicians should inspect to ensure they do not miss something. Additionally, there may be pests that are not included on the list, but most of those will be covered in the second half of the manual.

Small flies. Most fly species are easily recognized, but there are several small flies that require close examination. For example, close inspection will show the long antennae and weak flight of fungus gnats that differentiate them from phorid flies, and the wings of phorids that separate them from fruit flies. This is important because accurate identification can make the difference in selecting effective chemical and non-chemical control methods. For example, recognizing flesh flies on a light-trap glueboard may guide the service technician to inspect and find a malfunctioning air curtain, rather than spending time searching for an indoor breeding site. Recognizing moth flies and dark-eye fruit flies on the walls and ceiling could lead to a close inspection of floor drains for clogging organic matter.

Beetles. Holes in wood are created by a few species of insects, and these can be identified by the diameter of the holes and the type of wood, with proper identification requiring a ruler and distinguishing of hardwoods from softwoods. Beetles that infest structural timbers create galleries below the surface and exit holes in the surface. Thus,

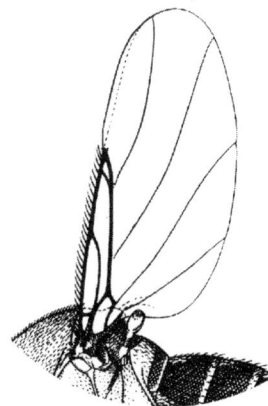

Phorid wing - close up

Lyctid Beetles
1/32 to 1/16 inch

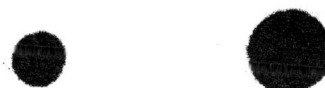

Bark Beetles
1/16 to 3/32 inch

Anobiid Beetles
1/16 to 1/8 inch

Beetle exit hole sizes

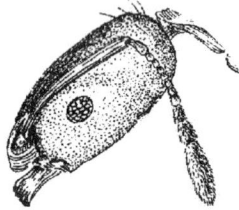

Pharaoh ant close-up

distinguishing the round or oval holes used by adult beetles to escape will enable the technician to identify the beetle that is infesting the wood. However, it is important to note that it won't tell if the infestation is current or old. Determining the diameter of the round holes and the wood in which they are found separates the Lyctid beetles (1/32 inch in hardwoods), Anobiid beetles (1/16 inch in softwoods), flat-headed borers (1/2-inch, flattened oval), and round-headed borers, (3/8-inch, rounded oval).

Ants. Identification of an infesting ant species often is critical to a successful control program, but ants are small, and magnification is usually necessary. Species often look alike, but are separated by small differences in the number of segments at the end of the antenna or the nodes between the thorax and abdomen. Foraging behavior and seasonal activity can be helpful: small ants found foraging throughout the house most likely are pharaoh ants; large black ants that are active indoors most likely are carpenter ants. However, thief ants also forage widely indoors and are the same size as pharaoh ants, but thief ants have a distinctly two-segmented club at the end of the antennae. Color and habits may not be enough to distinguish all the ants that occur around the perimeter and inside structures.

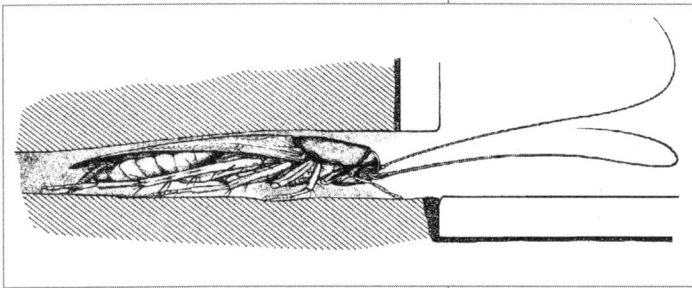

Cockroach in crevice

CONDITIONS
Insects, spiders, and rodent pests require a set of suitable conditions to survive; these include relative humidity or moisture, food, and harborage. These are basic requirements for survival and for populations to grow to infestation levels. The range and quality of these requirements vary: cluster flies survive in attics that have a relative humidity as low as 20%; adult fruit flies need at least four times that to fly, mate, and lay eggs. The quality of the food eaten by the larval stage of wood-infesting beetles can influence the adult size and the number of eggs laid by the female. The larval food of the fruit fly can determine where the female lays her eggs. Harborage is often overlooked as an important survival requirement. For many insects it is crucial, and potent aggregation pheromones are deposited there to maintain large numbers and attract males and females. German cockroaches deprived of suitable harborage do not reproduce or survive as long as those with adequate harborage.

Prevention. Inspection must have the added objective of identifying the source of the target pests and the conditions that permit the infestation or the points of access for these pests. Once the conditions providing the infestation are identified and changed or the entry point sealed (non-chemical methods), there can be long-term control. Without some level of prevention, control will be less effective and infestations will persist. Inspection has the opportunity to find and identify the pest and to determine immediate and lasting control methods.

STICKY TRAPS
The typical use for sticky traps placed indoors is to monitor or detect the presence of pests before and/or after chemical or non-chemical treatment. For pest management programs, sticky traps can direct chemical application to increase efficacy and reduce waste. However, there are other effective uses for sticky traps in residential and commercial pest control. When placed close to harborages, sticky traps can remove and reduce infestations of pests that occur in low numbers. House centipedes, field crickets, and wolf spiders can be removed more effectively with sticky traps than with insecticide application.

Sticky trap

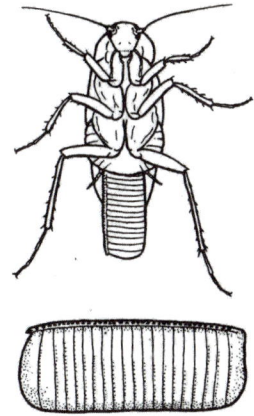

Careful inspection of the insects captured on sticky traps can provide information on the location of infested harborages. For example, the presence of female German cockroaches carrying egg cases is an indication that an infested harborage is close to the trap. These females usually remain inactive during the approximately 28 days they are carrying the egg case. They leave the harborage in search of food only every few days and do not travel far. The presence of egg cases and the small nymphs that have hatched from them is another indication that an infested harborage is close to the placement of the sticky trap. The abundance of males in a sticky trap may not indicate a nearby harborage, however, because males and large nymphs travel far from harborages in search of food.

Sticky traps that remain in place for long periods and have captured cockroaches, especially large species, such as American cockroaches, and other pests on them may attract house mice. The mice eat parts of insects trapped on the glue. Evidence of this will be remnants of only the legs and antennae of the insects.

Sticky traps can be dated and saved to provide evidence of declining or increasing pest numbers. Insects and spiders captured on the glue will remain intact for months, at least for the time necessary to use them to show service progress. Sealing the traps in plastic bags will prevent other insects or rodents from eating the captured specimens.

Cockroach showing egg case

UV LIGHT TRAPS

The insects captured by ultraviolet (UV) light traps can show the presence of pest infestations, particularly moths and some beetles. The monthly capture data can be used to track the life cycle of the pests, as indicated by the increase and decrease of number of adults captured. Because insects, such as click beetles, flesh flies, and sod webworm adults, also will be attracted to these lights, their presence on the glueboards will mean that there is an opening they used to enter, and may indicate that an air curtain is not adjusted or a door or window is not closed.

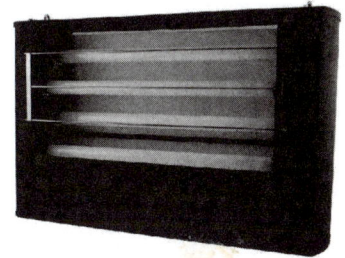

Light trap

OVERWINTERING PESTS

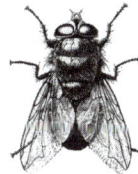

Cluster fly

- **Cluster flies** gather on the sunny sides of houses and other structures in late afternoon in August and September. They move through cracks and crevices and enter the attic, wall voids, or living spaces to spend the winter. **See page 77.**

- **Boxelder bugs** gather in large numbers along the sunny side of buildings, foundations, and the base of trees in fall. They form large aggregations of nymphs and adults that overwinter in protected locations. **See page 43.**

Boxelder bug

- **Stink bugs** gather as individuals on the sunny and/or warm sides of houses and other structures in fall. They crawl through cracks and crevices around doors, windows, and soffits to spend the winter indoors. **See page 42.**

Stink bug

- **Kudzu bugs** gather as individuals on the warm and/or sunny sides of houses and other structures. They overwinter as individuals in protected locations. **See page 44.**

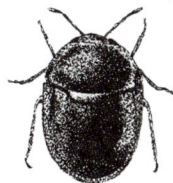

9

Western conifer seed bug

Asian ladybird beetle

Wasp

Umbrella wasp

Mosquito adult

Mud dauber nest

Pavement ant

Carpenter ant

Cluster fly

- **Western conifer seed bugs** come indoors in fall in small numbers. They may be under bark and come in with firewood. Once inside, the adults become active and fly; they make a buzzing sound in flight. They do not bite. **See page 45.**

- **Asian ladybug or ladybird beetles** gather in large numbers on the sunny and/or warm sides of houses and other structures in fall. They crawl through narrow openings to get inside the house or building; they sometimes overwinter in large masses outdoors. **See page 58.**

- **Elm leaf beetles** are associated with large, established elm trees; they often overwinter around the outside of buildings or in large numbers in the attics of houses. **See page 58.**

- **Yellowjackets** spend the winter in protected places, such as under loose bark of trees and firewood; they also overwinter in attics of houses that are close to their summer nest site. These are the queens that will found nests in spring. **See pages 39-40.**

- **Umbrella wasps** overwinter in attics, usually near the site of their summer nest. They are active in early winter (January) and fly to windows and lights in the living space. These are queens that will found nests in spring. **See page 37.**

- **Mosquitoes** sometimes overwinter as adults, especially the species that frequently breed around houses. These are females that select protected places to rest during winter; they usually do not bite during winter. **See page 80.**

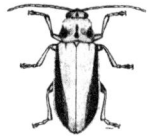

CRAWLSPACE AND ATTIC

- **Yellowjackets** build nests in attics, usually close to entry points, such as vents or louvers to the outside. The German yellowjacket commonly builds large nests in attics. **See pages 39-40.**

- **Mud daubers** build their mud nests close to attic openings, such as soffits and louvers at the end of gables, or close to the opening of a crawlspace. Adult mud daubers are active in spring; they are not aggressive and usually do not defend their nest. **See page 37.**

- **Pavement ants** build nests at the top of foundation walls, between the foundation and the sill plate. Their nests often contain wood shavings and can be mistaken for those of carpenter ants or wood-infesting beetles. **See page 30.**

- **Carpenter ants** build primary or satellite nests in moisture-damaged wood in attics, typically in rafters that are below leaks around vent pipes or damaged roof shingles. **See page 31.**

- **Cluster flies** will be in attics in fall and winter; they are generally not active during winter, and they leave the attic in spring. There may be accumulations of dead flies. **See page 77.**

or Paper wasp

- **Umbrella wasps** build nests in attics, usually close to entry points, such as vents or louvers to the outside. **See page 37.**

- **Pine sawyer** damage may be seen in attic rafters and framing, and also in exposed floor joists in crawlspaces. These are never active infestations. **See page 50.**

- **Anobiid powderpost beetle holes** may be seen in attic timbers, but they are most common in exposed floor joists in crawlspaces. Frass may spill from emergence holes well after the infestation has died out. **See page 48.**

- **Pine bark beetle damage** may be found in attics and crawlspace timber that has a small amount of bark remaining on the pieces. These beetles do not remain active in structural wood. **See page 49.**

- **Termite tubes** may be visible in crawlspaces; they typically are built against the foundation wall and extend from the ground to wood above, such as floor joists or the sill plate at the top of the foundation. **See page 90.**

- **Black widow spiders** build their webs close to doors and inside in corners near doors. The nests usually are close to the ground or floor. They usually do not occur in large numbers inside buildings. **See page 105.**

- **Camel crickets** occur in crawlspaces, usually when there is high humidity or water in the crawlspace. They use wood, paper, and other objects as harborage. **See page 68.**

OUTBUILDING

- **Black widow spiders** occur in crawlspaces, usually close to the entrance if it is at ground level. Cellar spiders also occur in crawlspaces, but they are harmless. **See page 105.**

- **House spiders** build webs between objects on shelves and equipment hanging next to walls. They can be very numerous in outbuildings, especially in late summer. **See page 102.**

- **Carpenter ants** that are active in or around outbuildings may be a sign that there is a satellite colony in the structure. Moisture-damaged wood at or above ground level may be supporting a nest of carpenter ants. **See page 31.**

- **Umbrella wasps** build their nests along the roof lines of outbuildings or sometimes at or inside doors. In such undisturbed areas, a nest can grow large by late summer and fall, and wasps will become aggressive to defend the nest site. **See page 37.**

- **Rats** build their nests close to the foundation of outbuildings. They often do so when there are animal pens nearby or when animal food, bird seed, and other organic material is stored in the building. **See pages 110-111.**

Umbrella wasp nest

Pine sawyer damage

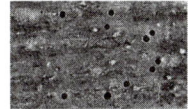

Anobiid beetle damage

Pine bark beetle damage

Termite tubes

Black widow spider

Camel cricket

Carpenter ant

Rat

Carpenter bee

Brown recluse spider

Termite damage

Black widow spider

Carpenter ant

Woods cockroach

Longhorned beetle larva

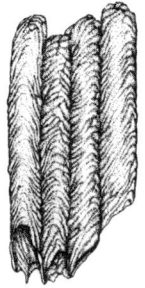

- **Mud daubers** build their mud nests close to door openings or inside if there is a clear access from the outside. The nests usually are built up high and may be unnoticed. The wasps are active in spring and may be seen collecting mud or water nearby. They are not aggressive. **See page 37.**

- **Carpenter bees** build their nests in spring. They select sites on sunlit sides of buildings, usually high off the ground. The entry holes in outbuildings may be hidden under eaves or along door frames. The females cut the entry hole and make the gallery; the males remain at the nest site and threaten anyone approaching the nest, however males cannot sting. **See page 36.**

- **Brown recluse spiders** build their webs inside buildings and often are found in clothing materials in outbuildings. There may be several of these spiders inhabiting an outbuilding. **See page 105.**

- **Subterranean termites** infest wood that is close to the ground or in contact with the ground. Infestations may be linked to a large colony that also is infesting a nearby house or to a separate colony. These infestations often go unnoticed. **See page 90.**

FIREWOOD

- **Black widow spiders** often occur around the edges of woodpiles; they may build webs in exposed sites between logs. In large wood piles there may be two or three females, but there will be a long distance between them. **See page 105.**

- **Carpenter ants** build nests in logs that contact or are close to the ground. There may be accumulations of fibrous frass in the end of some tunnels. Nests in wood piles may be satellites for a large colony nearby. Foraging trails may extend to the house. **See page 31.**

- **Woods cockroaches** may be found under the bark of hardwood and softwood logs; the nymphs are wingless and light brown. Adult males fly to lights at night in the fall. **See page 63.**

- **Subterranean termites** may infest logs that contact the ground; the workers and soldiers may be found at ground level or in soil-lined galleries in logs. Infestations in wood piles may not extend to nearby structures. **See page 90.**

- **Longhorned beetle larvae** may occur at the surface of or deep within logs that do not have contact with the ground; **flatheaded beetle** larvae usually occur under the bark of hardwood and softwood logs. Beetle larvae feeding under the bark of exposed logs can be heard several feet from the woodpile. **See pages 50, 52.**

WOOD-SHINGLE ROOF

- **Earwigs** are good climbers and flyers, they can move from tree branches that contact roofs or fly to the dark and moist habitat of weathered shakes. The shaded and north sides of houses may have the greatest numbers. **See page 69.**

- **Woods cockroaches** can find harborage and breed between damp and decaying shakes that are in areas not exposed to sunlight; these are favorable sites for these cockroaches. Wood cockroaches go toward lights at night. **See page 63.**

- **Carpenter ants** can establish satellite nests in the moisture-decayed shakes and structural timbers below the shake of the roof. Roof leaks that occur under shakes or along the flashing of chimneys or vent pipes can create favorable sites for these ants. **See page 31.**

- **Umbrella wasps** can find suitable nest sites at the bottom edges of shake roofs, and can be out of sight. These nests can house overwintering females in fall and winter. **See page 37.**

- **Silverfish** are common inhabitants of shake roofs because of the suitable harborage and available food. They also can occur in large numbers in the attics under these roofs. **See page 73.**

TRASH AND DEBRIS

- **The house mosquito and Asian tiger mosquito** breed in standing water that collects in trash and debris. They can produce hundreds of adult mosquitoes in a small amount of water. **See page 80.**

- **Earwigs** take advantage of narrow harborages found in accumulated debris around houses and outbuildings. They are good flyers and move from there to houses. **See page 69.**

- **Rats** may find little to eat in accumulated trash and debris, but they can use these as nest sites and move from there to forage in garages or indoors. Inspect for burrow entry points. **See pages 110-111.**

- **Deer mice** may nest in these sites or visit them in search of food. From these locations, mice can move to the perimeters of houses or other structures. **See page 112.**

- **Raccoons** are scavengers and often visit trash and debris sites in search of food, or they may nest there for short periods. **See page 116.**

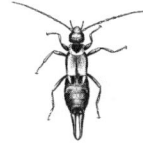

Earwig

Woods cockroach

Carpenter ant

Umbrella wasp nest

Silverfish

Mosquito adult

Rat

Deer mouse

Boxelder bug

Elm leaf beetle

Tent caterpillar

Aphid group on stem

Spots on leaves

Sweat bee nests

Cicada killer wasp

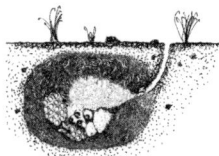

Bumble bee nest

TREES, PLANTS, TURFGRASS

- **Boxelder bugs** feed on the seeds of most maple trees (boxelder is a maple). The presence of boxelder or other maples trees in a neighborhood may supply food for these bugs. **See page 43.**

- **Elm leaf beetles** attack elm trees; the larvae eat the surface of the leaves. The adult beetles overwinter outside or inside (in attics). The pale green color of these beetles makes them easy to identify. **See page 58.**

- **Tent caterpillars** attack a variety of trees, but usually prefer wild cherry trees. The large "tents" are built in the spring. The caterpillars may be inside the tent during the day and leave to feed on the leaves at night. **See page 86.**

- **Aphids** are pests of almost all ornamental trees, shrubs, and other plantings. They are present throughout the spring and summer. They suck plant sap and produce honeydew on which ants feed. Controlling aphids will help control ants. **See page 43.**

- **Spots on leaves** and other discolorations of leaves on ornamental plants are probably not caused by insecticides. They most likely are caused by aphids or other insects that attack plants. **See page 44.**

- **Sweat bees** build their nests in exposed soil, often on banks that face the sun. There may be a large number of holes, and these small bees will be active during the heat of the day. They seem to be attracted to people working nearby. **See page 36.**

- **Cicada killer wasps** are active in late summer and fall, when the annual cicadas are singing in trees. This large wasp will dig holes in exposed soil or turfgrass, then capture a cicada to place in the hole. These wasps are not usually aggressive, and their activity is generally limited to a few weeks. **See page 37.**

- **Bumble bees** are active in spring when the overwintering queens look for a nest site. They often select old mouse and chipmunk burrows. Nests may be adjacent to houses and concealed among plants; nests also may be under decks. **See page 37.**

- **Yellowjackets** build below-ground nests in protected places, such as along the edge of a flower garden or in the open. The opening to the nest is often difficult to see. In fall, these nests can have a large number of wasps, and they can be aggressive. **See pages 39-40.**

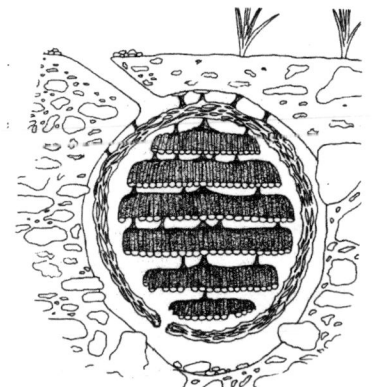

Yellowjacket nest below ground

- **Black Formica ants** are common around houses; they build low-profile mounds in turfgrass, usually on slopes facing the sun. These ants forage honeydew from aphids in trees and shrubs, and they occur on flowers. They can be mistaken for carpenter ants. **See page 35.**

- **Yellow ants** are common around structures; nests may be around the perimeter and along the foundation of buildings, and under concrete slabs. Colonies produce swarms in spring and fall, and the winged females can be confused with termite swarmers. **See page 30.**

- **Fire ants** are common in residential areas, and their mounds can be seen along the edge of turfgrass and sometimes close to structures. **See pages 33-34.**

- **Ticks** often are found along the edges of turfgrass, where there may be an edge with shrubs or a wooded area, and in areas of tall, uncut grass. The American dog tick and the lone star tick are common in these locations. **See page 96.**

- **Deer ticks** occur in areas that have white-tail deer populations and deer mice. The adult ticks feed on the large animals, but the nymphs feed on deer mice. This small tick spreads Lyme disease to people, dogs, and cats. **See page 97.**

- **Millipedes** live and feed in the thatch layer of turfgrass. Sometimes conditions result in the development of large populations. When the habitat becomes crowded, large numbers of millipedes migrate. **See page 107.**

- **Sod webworm moths** commonly are seen flying across turfgrass at sunset. These moths collect at outdoor lights and sometimes move indoors. They are distinguished from meal moths by their long, tubular shape. **See page 86.**

- **Bagworms** are caterpillars that live in a protective case as they feed on the leaves of trees, especially evergreens. The cocoons may remain on the tree or be found on the sides of buildings. **See page 87.**

- **Voles** are small rodents that feed on plant material in burrows below the ground. They make extensive runways; after snows melt, their above-ground tunnels can be seen. **See pages 111-112.**

- **Moles** are below-ground feeders on insects and earthworms; their feeding tunnels are visible in turfgrass in the spring and fall. **See pages 112-113.**

- **Squirrels** are a pest of turfgrass when they dig small holes in grass and flower beds in search of food, which includes nuts and bulbs. **See page 114.**

- **Skunks** forage at night and come to the perimeter of houses and other buildings in search of food scraps. They dig small holes in turfgrass searching for grubs below the surface; sometimes the damage to turf can be extensive if there is an infestation of beetle grubs. **See page 117.**

Black Formica ant

Fire ant

American dog tick

Deer tick nymph

Millipede

Sod webworm

Bagworm

Vole burrows & Vole

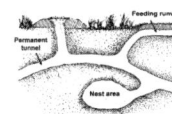

Mole burrows & Mole

Skunk

Carpenter bee

Carpenter ant

Skunk

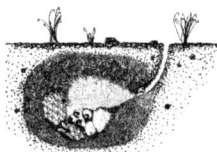

Bumble bee nest

Clover mite

Yellow ant

Pavement ant

Pharaoh ant

Chipmunk

WOOD DECK, PORCH

- **Carpenter bees** search for exposed, usually unpainted wood in which to build their nests. Females select flat surfaces and end-grain pieces of wood. They are active in the spring, and can be distinguished from bumble bees by their shiny abdomens. **See page 36.**

- **Carpenter ants and acrobat ants** build nests in exposed wood, especially wood that has been moisture damaged. Carpenter ants nesting in decking wood may forage away from the site and be unnoticed. They forage at night late in summer. Acrobat ant nests are small and may be detected by the powdery frass that falls from openings in the galleries. They both make chemical trails to their nests from foraging sites. **See page 31.**

- **Skunks** are active at night and generally are unnoticed except for the odor they leave behind. In spring, females may nest in protected sites under ground-level decks and porches. Sealing access to these areas will prevent this behavior. **See page 117.**

- **Bumble bees** often build nests at the edge of or under ground-level decks and porches; they select old mouse burrows as nest sites. The nests are difficult to locate when in these areas; unless they are threatening people, the nests may simply be left alone. **See page 37.**

AROUND FOUNDATION

- **Clover mites** are a problem in spring when large mite populations in turfgrass cause some of them to move onto foundation walls, and then up to windows and inside houses. These mites are common in newly established turfgrass. **See page 98.**

- **Yellow ants** occur around foundations; they often build their nests in the soil close to the foundation wall. Colonies produce swarmers (winged ants) in spring and fall, and they are sometimes confused with termite swarmers. **See page 30.**

- **Pavement ants** are common around foundations; they build their nests in the soil close to the foundation, and sometimes establish a nest between the sill plate and the top of the foundation wall. **See page 30.**

- **Odorous house ants** build nests around foundations of houses; they forage indoors and outdoors (on aphids). From these nests they may move permanently indoors. **See page 34.**

- **Chipmunks** may build ground nests along foundation walls and adjacent shrubs. They forage for food along the foundation, in garbage containers, and out into the surrounding areas. **See page 113.**

- **Deer mice** occur around foundations in fall when temperatures drop and the foundation acts as a heat sink to retain heat overnight. Once close to the foundation, the mice will look for entry points into the house or other buildings **See page 112.**

Deer mouse

- **Mason bees** often are seen around foundations and brick veneer on buildings. In spring, females look for small holes to enlarge and use for a nest site. They will tunnel into old mortar between bricks to build a singular nest. **See page 36.**

Mason bee

- **Wolf spiders** are common on the ground around foundations where they hunt for insects and other prey. They can move indoors through cracks and crevices around doors and windows. **See page 104.**

Wolf spider

- **Black widow spiders** can be found along the foundation, especially near the openings to crawlspaces, near crawlspace vents, and near downspouts. They are not usually numerous. **See page 105.**

Black widow spider

- **Field crickets** often are seen around foundations in fall when night-time temperatures begin to fall. The foundation retains warmth into the night and these crickets come to the heat source, and then move indoors. However, they can't survive indoors because it is too dry for them. **See page 69.**

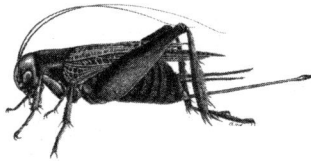

- **Sowbugs** are in the mulch and moist, decaying organic matter around foundations. They move inside when their habitat becomes too wet or too dry. They quickly die indoors because of the low humidity. **See page 99.**

Sowbug

MODERN LOG HOUSE

- **Ambrosia beetle galleries** are easily recognized by their dark blue or black staining. These beetles were active when the wood was freshly cut and its moisture was high. They do not remain active in seasoned wood. **See page 52.**

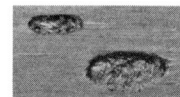
Ambrosia beetle damage

- **Pine sawyer galleries** are recognized by their oval shape and the presence of fibrous frass inside them. These beetles inhabited the wood when the tree was live or freshly cut; they do not re-infest seasoned wood. **See page 50.**

Pine sawyer holes

- **Buprestid-beetle or flat-headed-borer galleries** are distinguished by their flattened oval shapes. These beetles may be in freshly cut or standing timber. They sometimes occur in seasoned logs, but only rarely re-infest and do not cause structural damage. **See page 52.**

Buprestid beetle damage

- **Solitary bees** sometimes use the galleries of wood-infesting beetles as nest sites. They may excavate the gallery and remove old frass or produce new frass when they chew into the wood. This can give the false impression of an active beetle infestation. **See page 36.**

Anobiid beetle damage

Carpenter bee gallery

Carpenter ant

Acrobat ant

Woodpecker

Water strider

Backswimmer

Mosquito adult

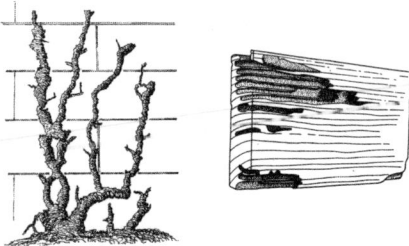

Termite tubes and galleries

- **Anobiid beetle exit holes** can be found in some logs used for modern log houses. Typically the infestation is not active, but there may be some frass falling from the holes, depending on where they are located. **See page 48.**

- **Carpenter bees** are attracted to the amount of exposed wood, especially the end grain of some logs. They often return to the same location year after year, and infestations can damage logs. **See page 36.**

- **Carpenter ants** are a major pest of log houses, especially when rainwater is not kept off the sides and away from the corners where logs meet. These ants will find and build nests in moisture-damaged wood. **See page 31.**

- **Acrobat ants** will take advantage of moisture-damaged wood on porches and decks to establish nests. They may go unnoticed until the powdery frass from their galleries is found. **See page 31.**

- **Woodpeckers** may be attracted to the sides of log houses for several reasons, one of which may be the presence of insects below the wood surface. The nests of carpenter bees contain many larvae that these birds can detect. **See page 118.**

SWIMMING POOLS

- **Water striders** occur on the surface of swimming pools in suburban areas. They can walk on water because of non-wettable hairs on their feet. The adults can fly long distances from natural ponds to pools. They do not infest swimming pools. **See page 44.**

- **Backswimmers** fly to swimming pools from nearby ponds and lakes; they are attracted to light at night. They swim underwater with long sweeping strokes of their back legs. These insects are known to bite people in swimming pools. **See page 45.**

- **Mosquitoes** will not live in swimming pools, but they may occur around pools and rest in the surrounding vegetation. **See page 80.**

TERMITE INFESTATIONS

- **Subterranean termite damage** to structural wood is indicated by the presence of galleries in wood lined with soil; typically the galleries follow the grain of the wood. Mud tubes may be found extending from soil to above-ground wood. There is no frass, such as fibrous pieces of wood in the galleries.

  The galleries may extend along the length of pieces of wood and to the wood above. When galleries follow moisture-damaged wood, they may extend into wall framing and to floors well above the foundation. In extended and severe infestations, a large portion of the colony may be located in wood above ground. **See pages 90-91.**

- **Drywood termite damage** is indicated by irregular galleries in the wood, typically the galleries do not follow the grain pattern of the wood. The colony is entirely above ground; there is no connection to soil by mud tubes. The sides of the galleries are not lined with soil, but are smooth. Colonies are generally small and isolated to a few pieces of wood.

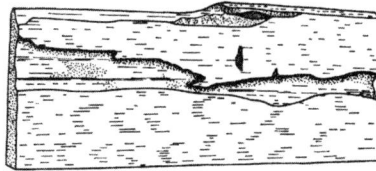

  The galleries of drywood may contain granular frass in the form of barrel-shaped pellets. The pellets are uniform in size and hardened, they can not be crushed by rubbing between fingers; they have distinct ridges. There are holes (called kick-out holes) along some galleries where the frass pellets have been ejected from the gallery. **See page 91.**

Drywood termite frass

Moth fly adult

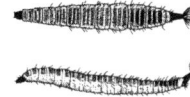

Moth fly larvae

FLOOR DRAINS

- **Moth fly** adults do not move far from their breeding sites; the presence of moth flies is an indication that there probably are floor drains that are partially clogged with organic matter. The larvae are deep in the clogging material, with the ends of their bodies in contact with the surface. They breathe through the posterior end and crawl out onto the surface of the drain when the clog is close to the top, such as in shower drains. Control of drain flies in a clogged drain must include removal of the organic matter. **See page 79.**

Dark-eye fruit fly

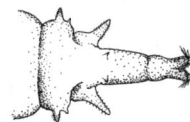

- **Dark-eye fruit fly** adults do not remain close to their breeding sites; they usually are found on walls and ceilings of the rooms that have infested drains. The larvae feed in the accumulated organic material within and at the top of the drain. They remain buried in the material when they feed, and only the posterior end of the abdomen is in contact with the air. They are capable of breathing through the opening at the end of the abdomen. They crawl away from the drain to pupate; the small, brown pupal case has extensions at the front. **See page 78.**

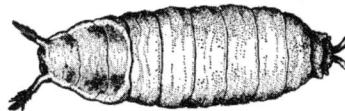

Fruit fly breathing tube

Fruit fly puparium

Phorid adult

- **Phorid flies** are sometimes observed coming through floor drains, or large numbers of adults may be seen around kitchen sink drains or other connections to the sewer system, whether it is a septic tank or standard sewer line. Adult phorids can be recognized by their erratic movement and wing veins. Phorid larvae may be feeding on organic matter around the top of the drain cover, but are probably feeding on organic material from a broken sewer pipe. The pupal case of phorids can be recognized by the boat-shape and the hornlike projections at the front. **See page 78.**

Phorid larva

Phorid puparium

- **American cockroaches** are associated with floor drains in the basements of buildings, especially with drains that are connected directly to the sewer line or the below-ground, storm-drainage pipes. These cockroaches commonly infest storm sewers in major cities, and American cockroach adults and nymphs can move from the sewer system through drain pipes that are not filled with water, and then enter basements. **See page 64.**

American cockroach

BATHROOM

Moth fly adult

Phorid adult

Carpenter ant

Carpet beetle larva

Sticky trap

Wolf spider

House centipede

German cockroach female

- **Moth flies** usually are seen on the wall close to the breeding site of the larvae. In a bathroom, larvae may breed in organic material around sinks and toilets, and in the clog of hair and debris in the shower drain. **See page 79.**

- **Phorid flies** can occur in bathrooms, but they don't usually breed in drains. The larvae of these flies feed in the rich organic matter associated with a broken or damaged sewer pipe. If a large number are found, check the sewer line or septic tank. **See page 78.**

- **Carpenter ants** usually are associated with moisture-damaged wood, which they prefer to use as a nest site. Check the wood around and under the bathtub or shower. Acrobat ants also may occur in this location; they also infest moisture-damaged wood. **See page 31.**

- **Psocids** are associated with high humidity conditions. These microscopic insects sometimes are found around fixtures on the sink or tub. They breed in organic material and mold on walls and cabinets. **See page 74.**

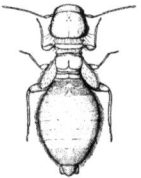

- **Carpet beetle larvae** can be found in bathroom closets and sometime climbing the wall. They feed on a variety of organic material, including leather and food scraps. **See pages 56-58.**

- **Rats and mice** sometimes occur in bathrooms; they can be attracted to moisture or have access through a crawlspace below. Mice can become trapped in bathtubs, as they are not able to climb up the smooth sides if they fall in. **See pages 110-111.**

- **Wolf spiders** may be found in bathrooms when the moisture and lights attract insects they can prey on. Place traps against walls in open areas. **See page 104.**

- **House centipedes** often are seen on bathroom walls and floor; they forage for insects and spiders. They can be captured in sticky traps placed in cabinets and against walls. **See page 106.**

- **German cockroaches** occur in bathrooms in houses or apartments that have a severe infestation. The bathroom may not offer food, but it provides harborage and humidity. If females carrying an egg case are trapped in the bathroom, it indicates there is a harborage nearby. **See page 62.**

- **Silverfish** are in bathrooms because of the humidity and the potential of finding food. They feed on microscopic mold and organic material in cracks and crevices. **See page 73.**

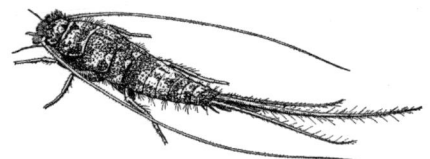

CLOTHING AND FABRIC

- **Holes in fabric** may indicate feeding by carpet beetle larvae, and clothes moth caterpillars. These insects attack natural fibers, such as wool and silk; they do not feed on cotton. Holes in cotton fabric actually may be caused by the teeth of the zipper on a garment hooking into a thread during washing. **See pages 55, 85.**

- **Carpet beetle larvae** may be on fabric or in closets or boxes that contain wool, silk, or furs. Sometimes only the cast skins of the larvae can be found along the edge of drawers or closets. **See pages 56-58.**

- **Carpet beetle adults** often are found in the glass bowl of light fixtures in rooms that have an infestation; this may be the only sign that these beetles are present. Check for natural fiber clothing, leather, and furs. **See pages 56-58.**

- **Clothes moth** adults usually are seen in closets or rooms that may have an infestation. These moths are small and do not naturally fly to lights. It is the caterpillar stage of the two clothes moth species that damage fabric and other organic material. **See page 85.**

- **Silverfish** eat protein-based material indoors. Their damage to fabric is usually associated with that which has food stains or food material in the weave of the material. The silverfish are eating this and damaging the fabric in the process. **See page 73.**

KITCHEN CABINETS

- **German cockroaches** occur in and around kitchen cabinets, under sinks, and in food-storage closets. Females carry their egg cases for about 28 days and deposit them in crevices in cabinets. **See page 62.**

- **Carpet beetle larvae** infest a variety of food materials, including spices and dry pet food. Evidence of an infestation may be cast larval skins on shelves. **See pages 56-58.**

- **Cigarette and drugstore beetles** are common in household food cabinets. Larvae feed on seeds, nuts, beans, spices, yeast, dried insects, fish, vegetables, flour, meal, and tobacco. The drugstore beetle also will attack leather and food-stained fabric. **See pages 52-53.**

- **Sawtoothed grain beetles** are a common pest of flour and cereal products; these are the small beetles found in boxes of noodles and flour. Adults and larvae can be found in cracks and crevices in cabinets. **See page 53.**

- **Rice weevils** are found infesting nuts; cereals and cereal products, such as macaroni and cake flour; and rice products. These are small beetles and may not be noticed until infestations are severe. **See page 54.**

Damaged cloth

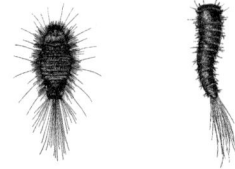
Carpet beetle larvae

Carpet beetle

Clothes moth adult

Silverfish

German cockroach female

Drugstore beetle

Sawtoothed grain beetle

Rice weevil

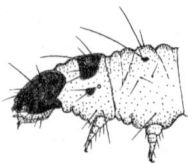

Indian meal moth caterpillar head

Indian meal moth

Psocid

Thief ant

House mouse feces

Carpet beetle larva

Carpet beetle

Bed bug adult

Wolf spider

- **Indian meal moth caterpillars** remain in the infested food material until full grown, then move away to pupate. Caterpillars can be found crawling on walls and ceilings away from the infested site. They are sometimes thought to be maggots when they are crawling on kitchen counters. **See page 84.**

- **Indian meal moth adults** usually remain close to the infested site, but also may be found in locations far from the kitchen. These distinctly marked moths remain inactive during the day and typically fly in early evening. **See page 84.**

- **Psocids** are found in cereal products and flour. They are almost microscopic and may be unnoticed until an infestation is severe. They can be found in cracks and crevices in cabinets and sometimes on counters. **See page 74.**

- **Pharaoh ants and thief ants** are about the same size and shape, and they infest similar materials. Both species forage widely indoors. Pharaoh ants have numerous small or satellite colonies scattered through a structure; thief ants have similar habits. Both species feed on high-protein materials in kitchens and elsewhere. **See page 32.**

- **Mouse droppings** may be found in cabinets and on kitchen counters. In fall and early winter, these droppings may be from deer mice; during other times of the year, the droppings may be from house mice. **See page 112.**

BEDROOM

- **Carpet beetle larvae** or the cast skins of larvae are commonly found in bedroom closets and dresser drawers. These larvae feed on a variety of organic material, including leather and natural fibers. Larvae can molt many times which can give the impression of a severe infestation. **See pages 56-58.**

- **Carpet beetle adults** are common in spring; at other times of the year, the adults are not active. When large numbers of adults are found in light fixtures or other areas, it indicates that mice have cached food, and it is infested with these beetles. **See pages 56-58.**

- **Bed bugs** can be detected with sticky traps. Adults and nymphs will crawl into traps that are placed along baseboards, under beds, and adjacent to bedside furniture. **See page 42.**

- **Wolf spiders** are active at night when they search for insects and spiders as food. They can be captured in sticky traps placed along baseboards and under beds and dressers. **See page 104.**

- **House centipedes** often are found in bedrooms where they are searching for insects and spiders as food. They are primarily nocturnal, and they move quickly on walls and ceilings. **See page 106.**

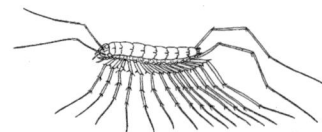

- **House spiders** build webs in corners close to the floor and will move along baseboards, beneath beds, and in closets. They frequently are captured on sticky traps placed in corners. **See page 102.**

BASEMENT

- **Pine sawyer** damage may be seen in attic rafters and framing, and also in exposed floor joists in crawlspaces. These are never active infestations. **See page 50.**

- **Anobiid powderpost beetle holes** may be seen in attic timbers, but they are most common in exposed floor joists in crawlspaces. Frass may spill from emergence holes well after the infestation has died out. **See page 48.**

- **Pine bark beetle damage** may be found in attics and crawlspace timber that has a small amount of bark remaining on the wood. These infestations do not remain active in structural wood. **See page 49.**

- **Termite tubes** may be visible in crawlspaces; they are typically against the foundation wall and extend from the ground to wood above ground, such as floor joists or the sill plate at the top of the foundation. **See page 90.**

- **Cellar spiders** often occur in corners or between exposed floor joists. These can be large spiders with unorganized webs; they are harmless. **See page 103.**

- **Woodlouse spiders** are common in the fall; they often are trapped in sticky traps placed along walls. These spiders enter under doors and around basement windows from mulch and turfgrass around the foundation. **See page 103.**

- **Springtails** require humid conditions; they can be numerous in basements that have moisture problems. They can occur near doors that open to the outside and in shaded or dark corners of basements. **See page 67.**

- **Carpet beetle adults** may suddenly occur in large numbers, or be seen at lights or in light fixtures. This may indicate that a stored food material is infested, or that mice have cached food which has become infested. **See pages 56-58.**

- **Pavement ants** sometimes build nests at the top of foundation walls, beneath the wood sill. Debris from the nest includes pieces of wood and has the appearance of that of a wood-infesting beetle. **See page 30.**

- **Pillbugs and sowbugs** may be seen dead and curled up on the floor or crawling in moist locations. They have come in from the mulch and vegetation surrounding the basement. **See page 99.**

- **Deer mice and house mice** will build nests in basements; inspect corners where floor joists provide a protective location. **See page 112.**

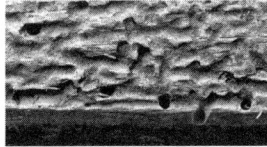

Pine sawyer damage

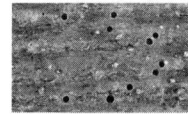

Anobiid beetle damage

Termite tubes

Cellar spider

Woodlouse spider

Springtail

Carpet beetle

Pavement ant

Sowbug

Deer mouse

House mouse feces

Oriental cockroach

Wolf spider

House centipede

Field cricket

Camel cricket

Blow fly

House fly

Umbrella wasp nest

Dark-eye fruit fly

- **Mouse droppings or feces** may be found along walls, on shelves, on the top of the foundation wall if it is exposed, on attic rafters and framing, and in exposed floor joists in crawlspaces. These do not necessarily indicate an active infestation. **See page 112.**

- **Oriental cockroaches** occur in the basements of commercial buildings and large apartment buildings in urban areas. They may be found outside the building perimeter during summer. **See page 63.**

- **American cockroaches** infest the basements of buildings; the infestation may be linked to drain pipes that connect to the sewer system. Set sticky traps near drains. If captured cockroaches are destroyed in the trap, this may have been done by mice. **See page 64.**

- **Wolf spiders** may be seen almost any time of year, but are most common in spring and fall. They come in from the perimeter where they live in mulch and vegetation. **See page 104.**

- **House centipedes** can survive in basements because of the availability of food, such as spiders and insects. Infestations are rarely large; sticky traps may be able to control them. **See page 106.**

- **Field crickets** occur in small numbers in fall when they move to the perimeter of buildings for heat on cold nights. Sticky traps can be used to remove the few that get inside. **See page 69.**

- **Camel crickets** may be found in dark basements that have a partial dirt-floor crawlspace or high humidity. They can be captured in sticky traps with cockroach-bait tablets. **See page 68.**

Dumpster

- **Blow flies** are attracted to the odor of garbage in Dumpsters; they will be active during the hottest part of the day. These flies follow food odors into restaurants near the Dumpster. **See page 76.**

- **House flies** are attracted to the odor of garbage, and females will lay eggs on exposed surfaces. Infestations can increase rapidly in summer when temperatures are high; adults will live about one month, which results in large numbers in fall. **See page 76.**

- **Umbrella wasps** often build nests in Dumpster stalls because of the protection; the availability of prey such as flies, and the presence of liquid food. These wasps can be aggressive in fall when the nests are large and there are many individuals. **See page 37.**

- **Red-eye fruit flies** occur around Dumpsters when there is an abundance of fruit and vegetable material. The maggots can survive in the humid conditions inside the Dumpster, and the adults can fly short distances to kitchen doors. **See page 77.**

- **Norway rats** are common around Dumpsters and Dumpster stalls that are not kept clean. Rats may not be nesting at these sites, but simply coming there to feed at night. They move from the Dumpster to buildings close to the stall. Rat droppings may be visible around the Dumpster or corners of the stall. **See page 110.**

OFFICE – STICKY TRAP

- **Pharaoh ants and thief ants** occur indoors throughout the year; these small ants build small nests in many locations in a facility. They will be captured in small numbers on sticky traps, but this is not an indication of a small infestation. **See page 32.**

- **Wolf spiders** can be caught in sticky traps in spring and fall, especially traps placed near doors that open to the outside. In severe infestations, a large number of spiders can be captured. Place traps along baseboards. **See page 104.**

- **German cockroaches** occur in small kitchens, employee break rooms, and areas where employees keep personal items. This cockroach can be brought into the facility from outside in food material and personal belongings. **See page 62.**

- **House centipedes** are captured on sticky traps that are placed along baseboards. Infestations usually are only a few individuals and sticky traps can help remove them. **See page 106.**

- **Silverfish** occur in most office environments, but they are rarely present in large numbers. Sticky traps placed where the silverfish have been seen will capture some of them. **See page 73.**

- **Bed bugs** may occur in office environments because they can be brought from infested homes or other buildings. Sticky traps placed where bed bugs have been reported or where bites have occurred can help to locate infested harborages. **See page 42.**

KITCHEN – LIGHT TRAP

- **Blow flies** are common during the warm months, especially if there is a Dumpster or garbage cans close to doors or windows leading to the kitchen. They fly to overhead lights during the day, but move to light traps at night. **See page 76.**

Pharaoh & Thief ant

Wolf spider

German cockroach female

House centipede

Silverfish

Bed bug adult

Blow fly

Dark-eye fruit fly

House fly

Phorid fly

Moth fly adult

Flesh fly

Fungus gnat

Indian meal moth

Sod webworm

Meal moth

- **Fruit flies** are common in kitchens; the red-eye fruit fly will be around fruits and vegetables, the dark-eye on walls and ceilings. When infestations are high, these flies will be caught in light traps. **See page 77.**

- **House flies** are attracted to the odor of food in kitchens and will enter through doors and windows. Flies move to light traps placed away from windows. UV light traps remove most house flies from indoors, but there may be some that will not move to the trap. **See page 76.**

- **Phorid flies** usually are not caught in light traps in large numbers, but females are attracted to UV light. The presence of these flies indicates accumulations of rich organic matter; sewer lines should be inspected for breaks. **See page 78.**

- **Moth flies** usually do not move far from their breeding site; the presence of moth flies in a light trap is an indication that there are floor drains that are partially clogged with organic matter. **See page 79.**

- **Flesh flies** are attracted to the odor of garbage in Dumpsters, and they can move from the Dumpster to nearby kitchens. The presence of these flies in light traps may indicate that an air curtain is not working properly, or there are doors or windows open. **See page 79.**

- **Fungus gnats** often are collected in light traps, sometimes in large numbers. They can move easily through window screens. They are not likely to breed indoors, but check potted plants for possible infestations. **See page 78.**

- **Indian meal moth adults** in light traps indicate that there is an infestation or infested material in the storage room. Light traps placed close to doors or windows may attract these moths from outdoors. **See page 84.**

- **Sod webworm adults** are attracted to UV light and often will be captured in light traps. They often are at outdoor lights at night and will move from there indoors. They are recognized easily by the rounded shape of the body and wings and the long palps that extend in front of the head. **See page 86.**

- **Midges** may be attracted to UV lights at night and may occur in light traps that are placed close to doors or windows. They are easily recognized by their large and bushy antennae and long legs. **See page 80.**

FOOD WAREHOUSE – LIGHT TRAP

- **Meal moth adults** in light traps indicate an infestation that may be far from the light trap. When there is an abundance of moths, there may be few larvae; when moth numbers decline in the traps, the larvae will be easier to find. Use pheromone traps when adult numbers in light traps are high. **See page 84.**

- **Indian meal moth adults** may be numerous in light traps during the warm months, when infestations often increase. Use pheromone traps as a control strategy or to locate the infestation when there are large number of adults; when larvae are developing, the adult numbers will be low. **See page 84.**

- **Flesh flies and blow flies** occur indoors during the warm months; this usually is an indication that there is a breeding source outside, such as garbage, and these flies are coming in through doors, windows, or an air-curtain that is not properly directed. **See pages 76, 79.**

- **House flies** occur indoors during the warm months; this is usually an indication that there is a breeding source outside, such as garbage; these flies usually come in through doors, windows, or an air-curtain that is not properly directed. **See page 76.**

- **Phorid flies** may be captured in light traps in small numbers; these flies are common and a few in a trap is normal. If the numbers are high, the facility may have decaying material, a damaged sewer pipe, or clogged floor drain. **See page 78.**

FOOD WAREHOUSE – STICKY TRAP

- **Larder beetles** will crawl away from an infested location and be captured in sticky traps. To pinpoint the site of the infested material, use a series of traps. **See page 57.**

- **American cockroaches** are common in large facilities. They are active outdoors during the warm months, and then move indoors during the winter. Sticky traps can help locate the infested harborages, but the adults will forage far from a harborage each night. **See page 64.**

- **Flour beetles** such as the sawtoothed grain beetle move away from an infested location when the population becomes large. Set a series of sticky trap or pheromone traps to locate the infested material. **See page 54.**

- **Spider beetles** infest a variety of grain and meal products; they can become numerous when infestations persist for a long time. These beetles usually do not fly; sticky traps with large numbers of these beetles may be close to the infested material. **See page 53.**

- **Pharaoh ants and thief ants** may occur in large warehouses; there are numerous places for the small nests or satellite nests produced by this species. Sticky traps are helpful in locating a nearby nest, but the best control is use of baits, not liquid insecticide. **See page 32.**

Flesh fly & Blow fly

House fly

Phorid adult

Larder beetle adult

Sawtoothed grain beetle

Spider beetle

Pharaoh & Thief ant

Wolf spider

Field cricket

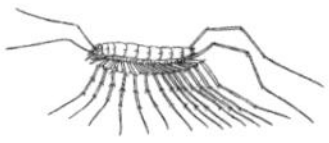

House centipede

Silverfish

CLASSROOM STICKY TRAP

- **Pharaoh ants and thief ants** occur in classrooms because of the opportunity to find nest sites and the potential of food scraps. Both species will forage over a wide area and form small nests. Baits are the most effective control strategy. **See page 32.**

- **Wolf spiders** move into buildings during spring and fall when they are most actively looking for a mate and to get out of the cold. Sticky traps along baseboards and near doors will help to eliminate many of these spiders. **See page 104.**

- **Field crickets** occur indoors in fall when the temperature begins to drop and the nights get cold. They enter around doors and windows. Sticky traps in dark corners will help to eliminate crickets. **See page 69.**

- **House centipedes** may occur in classrooms because of the availability of food, such as spiders and insects. They are not a threat to people unless they are handled, and they are difficult to catch. These centipedes are rarely numerous, but sticky traps can help remove them. **See page 106.**

- **Silverfish** can survive in a classroom environment because of the available harborages and the abundance of food scraps. Populations can be reduced with the use of sticky traps placed along baseboards and in cabinets. **See page 73.**

ADDITIONAL PESTS

Use these lines to list any pests you add to the chapters and the page numbers on which you add them.

Pest   **Page #**

_____   _____

_____   _____

_____   _____

_____   _____

_____   _____

_____   _____

_____   _____

_____   _____

_____   _____

_____   _____

CHAPTER 3

ANTS, BEES, AND WASPS

Ants, bees, and wasps have a range of habits: some are solitary, some live in large colonies, some are plant feeders, and some are parasites or predators. They have chewing mouthparts, and they use their mandibles to chew wood and build "paper" nests. Some species, such as bees, can lap water and nectar from flowers.

Ants, bees, and wasps build nests that contain their colonies. The colony survives through a system of dividing the labor of building and maintenance among members. Queens originate the colony and lay eggs; workers, which are sterile females, gather food, maintain and repair, and defend the colony.

Ant colonies usually start with a mating flight: males and females fly from the nest and mate in the air or on the ground. After mating, the female forms a nest by making a brood chamber, and begins laying eggs. Workers of the first brood forage and feed the queen, expand the nest, and care for the next brood. The founding queen continues to lay eggs and remains in the nest. When the colony reaches a size of several thousand workers, the queen lays eggs that develop into reproductive females and males, and the process starts over.

Bees and wasps often build nests in the soil at the perimeter of structures or in turfgrass. Sweat bees are commonly seen in spring as they build nests in exposed soil. Cicada killer wasps are active in fall when they are excavating holes in the ground to hold the cicadas they have captured as food for their larvae.

Yellowjackets build nests in the ground as large as the more-familiar nests in trees and shrubs. Below-ground nests start with a single queen excavating an abandoned mouse burrow. Once the first and second brood of workers develop, the nest expands with more and more brood cells. By late summer, the nest may have several thousand workers.

The venom of wasps, yellowjackets, and bees are similar, with all containing enzymes that cause swelling and redness. A bee sting produces a slight swelling and itching for several days. Further stings may produce no reaction in persons that become desensitized to the venom. However, others become more allergic with successive stings and reach a sensitivity level in which another sting results in an acute reaction or death.

ANTS

Ants are the most abundant insects on earth. Although there are thousands of species, relatively few are pests—indoors or outside. Most of those that are pests have specific habits or infestation sites: carpenter ants build nests in structural wood; pharaoh ants and thief ants live almost exclusively indoors; odorous house ants nest outdoors but forage indoors; and fire ants usually nest outdoors in large colonies.

Most ant colonies produce winged adults at least once a year, usually in spring and summer. Large numbers of males and females fly from the nest and collect on the outside of buildings. Some swarms occur indoors. Subterranean termites have the same swarming schedule. Termites usually swarm in spring, on warm days following a rain. Because they are about the same size and color, swarming ants are sometimes mistaken for termites. For example, the swarmers of the Southeastern subterranean termite (*Reticulitermes hageni*) are relatively small and can be confused with the swarmers of

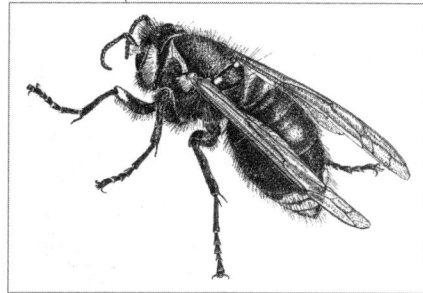

Wasp

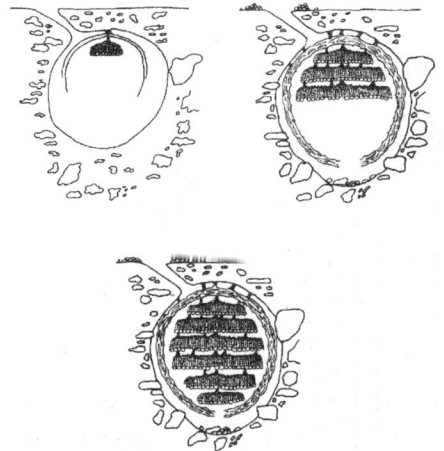

Underground yellowjacket nest development

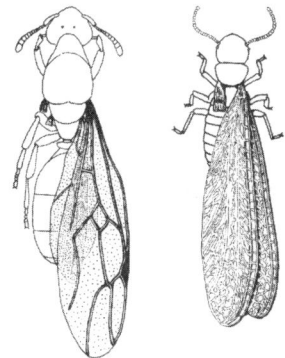

Winged Ant   Winged Termite

Ant vs Termite

Crazy ant

Larger yellow ant

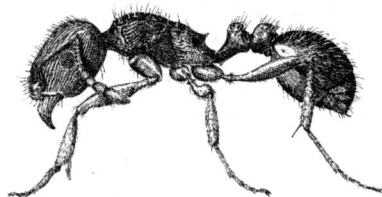
Pavement ant

the larger yellow ant (*Acanthomyops interjectus*). The swarmers of both are light brown, and emerge from colonies in the fall. (*Termites are discussed in Chapter 11.*)

The wings and body shape of ants and termites are distinctly different, however. Ants have two pairs of wings with one shorter than the other; termites have two pairs of wings that are both the same length. The antennae of ants are large and elbowed, the antennae of termites are small and not bent in the middle.

CRAZY ANT

The body color of the worker ant is dark brown to blackish brown. It is very slender, and the legs are extremely long.

Nests indoors often are built in wall voids. Outdoor nests are around buildings and sites such as trash cans and refuse Dumpsters. They occur on the ground and upper floors of commercial buildings and hospitals. The long legs of these ants make their walking seem unorganized, which is the origin of their name: crazy ants.

LARGER YELLOW ANT

Worker body color ranges from uniform yellowish brown to dark brown. The swarming adults have the same body color, and their wings are light brown.

Nests are in exposed soil or under the cover of stones or logs; in open areas, the nests are sometimes in small mounds. Nests may be around the perimeter of buildings, along the foundation, and under concrete slabs.

Colony. Nest construction and foraging is done primarily at night. Winged males and females usually emerge from mid-March to July, but some swarms can occur as late as September. Indoors, swarming may occur from late fall to early spring. The swarms of this ant are often confused with those of subterranean termites, because of their light brown color and spring and fall swarming habits.

PAVEMENT ANT

Worker body color ranges from light brown to blackish brown; the legs are light brown. The surface of the head and thorax has distinct longitudinal grooves, which help in its identification.

Nests are in exposed soil, under stones and pavement, and along sidewalks. Indoor nests may be in masonry walls and along the foundation.

Colony. These ants have large colonies, and there is usually one functional queen in a colony. Winged males and females swarm in June and July, but they may emerge almost any time of the year.

Habits. Natural food includes live and dead insects, honeydew, plant sap, and seeds. Indoors they forage for meat and grease. Nests are sometimes positioned between the wood sill and the top of the foundation wall, whether it is block or poured concrete. Refuse expelled from the nest site includes fragments of seeds, dead insects, and fine wood fibers. The presence of wood fibers can be alarming, but these ants do not infest wood.

WESTERN BLACK CARPENTER ANT

Worker body color is black, and there are strong hairs on the body. The legs are usually dark red.

Nests are established in damp or decaying wood; mature colonies contain 9,000 to 50,000 workers.

Colony. Egg laying is from April to June. Larval development is during the summer, with most of the workers produced by October. Egg production stops by August and September, and the larvae present in the colony overwinter with the queen. No food is consumed by the colony from October through January.

Habits. This is the dominant carpenter ant in the northwestern U.S. and adjacent areas in Canada. The satellite colonies of this species can be large, and from these they can move into and infest a nearby structure. Damage to structural wood can occur in a short time and is not limited to moisture-damaged wood. Satellite colonies in houses can contain several thousand ants. These ants typically maintain trails between the main colony and the satellites.

EASTERN BLACK CARPENTER ANT

The body color of the workers is typically black, though some may be reddish black. The large size of the queen and workers makes them easy to identify.

Nests are in dead trees, rotting logs, and stumps. Indoor nests are in moisture-damaged and sound wood.

Colony. Activity begins in April and extends to November. In some regions, they are inactive from December to April. Colonies produce swarms at three to five years of age.

Habits. These ants forage primarily at night, usually peaking soon after sunset, extending into the night, then peaking again before sunrise. Well-established trails connect satellite nests and the main colony nest site.

ACROBAT ANTS

Workers are yellowish brown to blackish brown. These small ants are identified by their heart-shaped abdomen.

Nests indoors are in structural wood exposed to moisture: roofing, siding, and porches, as well as door and window frames.

Colonies. The colonies usually are small and contain 2,000 to 3,000 workers. Winged males and females emerge from early June to November.

Habits. The common name, acrobat ant, is derived from its habit of raising its abdomen over its head and thorax when disturbed. Workers also will bite and give off an odor when alarmed. Infestations may be discovered by the powdery frass that falls from the nest galleries. These ants will make trails from their outdoor foraging sites to the nest site in wood, especially in enclosed decks and porches with wood frames.

Western black carpenter ant

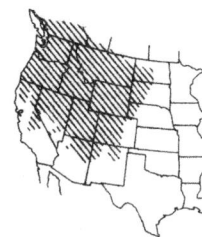

Western black carpenter ant distribution

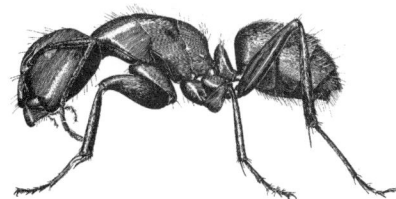

Eastern black carpenter ant

Eastern black carpenter ant distribution

Acrobat ant

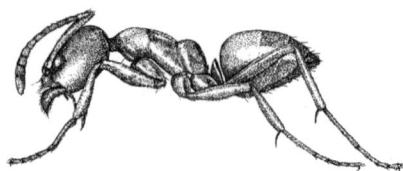

Argentine ant

Pharaoh ant

Thief ant

ARGENTINE ANT

Workers are light brown or brown.

Nests are in refuse piles, bird nests, wall voids, masonry voids, and cracks in concrete slabs around the perimeter of buildings. Nests also may be made deep in the ground during dry or cold weather.

Colony. The colonies of these ants usually are large and contain hundreds of queens. New colonies are formed when a queen and a small number of workers migrate to a new site. In winter, several colonies may combine to form a large colony.

Habits. Argentine ants produce a chemical trail, which allows them to forage day and night. Food includes sweets, meats, fruit, eggs, dairy products, animal fats, and vegetable oils. Excessively dry or wet conditions often cause workers to invade houses. Colonies can become dominant in an area and effectively force out other ant species, such as odorous house ants.

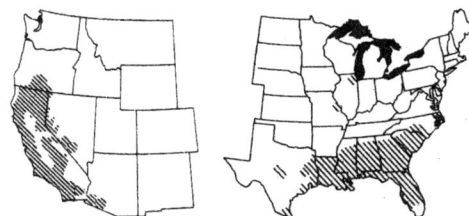

Argentine ant distribution west and east

PHARAOH ANT

Workers are yellowish brown to reddish brown. The segments of antennal club gradually increase in size to the end.

Nests indoors are in small secluded locations, such as in light switches, behind baseboards, and in cabinets. The nest sites are frequently moved.

Colony. These ants are active all year, and colonies can be large. New colonies are formed when a young queen and a small number of workers split from the parent colony.

Habits. Workers forage 24 hours a day and use chemical trails. Indoors, they feed on sweets, meat, grease, and a variety of other materials.

THIEF ANT

Worker body is shiny, yellowish brown to dark brown. The very last segment of the antenna is large and elongate.

Nests usually are in secluded voids in wood, masonry, and household materials.

Colony. Swarms occur from July to October. Colonies contain several hundred to several thousand individuals. Winged forms emerge from July to October.

Habits. Indoor food includes meat, sweets, ripened fruit, oils, and dairy products. This ant prefers food with high protein content.

LITTLE BLACK ANT

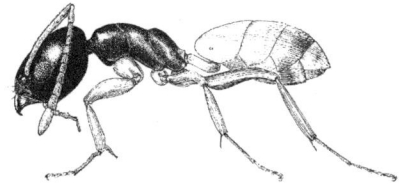

Worker body is shiny, dark brown to black. Each antenna has a three-segmented club.

Nests indoors are in structural wood or in the masonry of the foundation. Winged forms emerge from June to August.

Habits. Indoor food includes meats, sweets, bread, grease, oils, cereals, and fruit juices.

Little black ant

GHOST ANT

Head and thorax are brown; abdomen is yellowish brown; antennae and legs are pale brown.

Nests indoors are in voids and cavities, such as in closets and discarded clothing.

Colony. Colonies contain several hundred workers and several queens. Food is primarily sweets. Nests in southern regions are outdoors, in northern regions nests are indoors. They have been found infesting buildings much like pharaoh ants.

Ghost ant

WHITE-FOOTED ANT

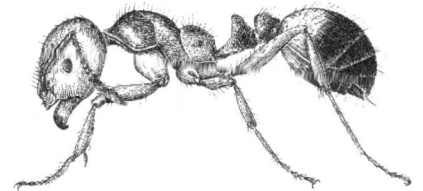

Worker body is blackish brown to black, and the tarsi are pale yellow.

Nests indoors are in wall voids, potted plants, and household materials; there may be several satellite nests. Trails are made outside buildings and indoors along the edges of baseboards and carpeting.

Colony. Winged females emerge in May and June. Indoors, they feed on sweets.

White-footed ant

RED IMPORTED FIRE ANT

Workers are yellowish brown to blackish brown.

Nests are above-ground mounds. Workers enter and exit through holes six to 10 feet from the mound.

Colony. Swarms occur from April to November, colonies produce 4,000 to 6,000 winged adults. Food includes meat, grease, and other protein-rich food.

Habits. When colonies are disturbed, workers become aggressive and deliver a painful sting.

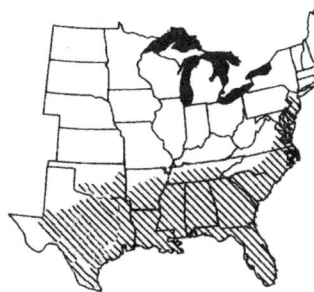

Red imported fire ant

Red imported fire ant distribution

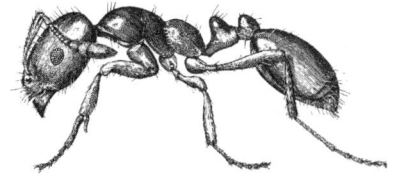

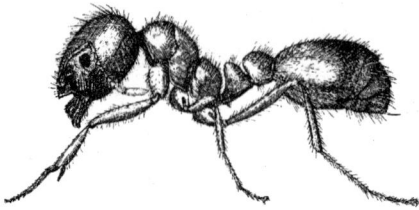

Southern fire ant

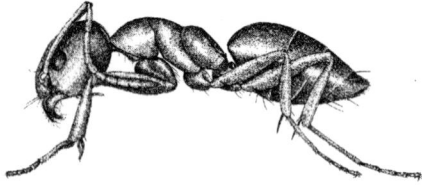

Odorous house ant

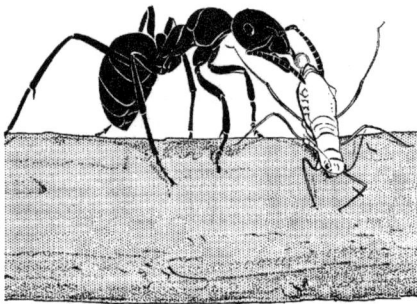

Ant with aphid

SOUTHERN FIRE ANT

The worker has a reddish-yellow head and thorax, and the abdomen is dark brown.

Nests are irregular mounds in loose soil, under stones, or in the masonry of houses. Nests can cover an area of several square feet.

Colony. Swarms occur from April to October, colonies are usually large. Food includes meat, grease, and oily nuts.

Habits. This native fire ant will become aggressive and sting when the nest is disturbed.

ODOROUS HOUSE ANT

Body color is uniform brown to black. Workers are less than 1/8-inch long. They are sometimes difficult to distinguish from Argentine ants, but the odorous house ant emits the smell of coconut when crushed.

Nests indoors are numerous and scattered throughout the structure, but are usually associated with moisture, such as in wall voids near water pipes, and in termite-damaged wood. A typical outdoor nest consists of a main colony and several satellite colonies, each with a queen and brood.

Colony. Each colony may have 200 functional queens; swarmers are produced in colonies that are four to five years old. Workers establish trails leading from the nest to food, and follow along tree limbs, the edges of buildings, baseboards, and kitchen counter tops.

Habits. When a nest is disturbed, workers run rapidly, emitting an odor from their elevated abdomens; they also can bite. Outdoors they feed on honeydew from aphids located close to the structure. Indoor nests usually are found near a moisture source; typical nests sites are in wall voids, behind paneling and baseboards, near water heaters, and along wall-to-wall carpeting. The indoor satellite nests are relatively small and may be moved frequently.

MOUND ANTS

Formica species ants are relatively large.

Nests. They build nest mounds that may extend only slightly above the surface of the soil, or be several feet above the ground, as does the Allegheny mound ant.

Colonies and the soil mounds they occupy can become large.

Habits. These ants forage during the day on plants and trees; they feed on insects and take honeydew from aphids.

ALLEGHENY MOUND ANT

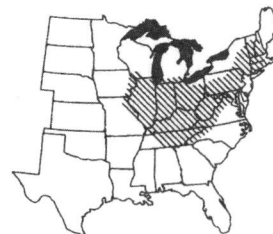

The head and thorax are reddish orange, the abdomen and legs are brownish black.

Nests are large mounds in wooded areas or in suburban turfgrass. Several mounds may be connected by tunnels that may extend into the ground and upward in the mound. Vegetation, including small trees and shrubs, is killed up to about 50 feet from the mound.

Colony. Allegheny mound ant colonies have multiple queens.

Habits. When a nest is disturbed, workers swarm out in large numbers and will aggressively bite.

Allegheny mound ant

Allegheny mound ant distribution

BLACK FORMICA ANT

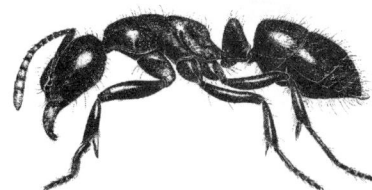

The body and legs are black, the surface is not shiny.

Nests are small mounds in the soil in open areas. Workers forage on trees and shrubs close to and far from the nest.

Colony. Colonies of this species typically have a single queen.

Habits. Workers will swarm out of a disturbed nest, but they generally are not aggressive. They are common in spring as soon as flowers emerge and aphids become active. Workers are common on spring flowers, especially peonies. They often are misidentified as carpenter ants, but they generally are smaller than carpenter ants.

Black formica ant

ASIAN NEEDLE ANT

Body is brownish black, legs and antennal segments are brown; the node is large. The eyes are small; the mandibles are large and light brown.

Nests are in dark, damp locations in soil beneath stones, logs, mulch, railroad ties, ornamental stones, and concrete pavers. The colony size ranges from a small nest of about 50 ants to large colonies of 50,000. Large colonies may have multiple queens. Workers are active beginning in March and continuing to October. Swarming occurs in July and August.

Habits. These ants can deliver a painful sting, similar to a fire ant. They forage on other insects, including termites. Their distribution may continue to spread.

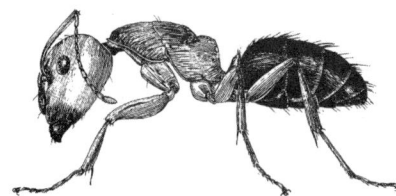

Asian needle ant

Carpenter bee

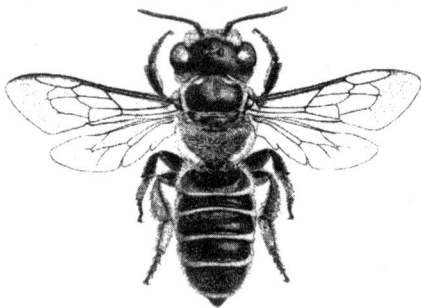

Sweat bee

Mason bee

BEES

CARPENTER BEES

These are large black and yellow bees that look like bumble bees. However, the abdomen of the carpenter bee is shiny; the bumble bee's is covered with fine hair.

Nests are built in exposed wood, including house siding or decking. Adults overwinter in the nest or in protected locations outdoors. Mating occurs in spring. Females may use an old gallery, create a new gallery, or adapt a gallery from an entrance used by other females.

Colony. Females make a half-inch diameter hole. After excavating the long gallery, they begin preparing individual cells. Pollen and nectar are placed together, and the female deposits an egg on it. Over several days, she will create about six cells in the gallery. Larval development is complete in 15 days, and adults appear in 36 days. The first bee to become an adult is usually in the cell at the end of the gallery. The adult cuts through the partitions of all the cells to emerge.

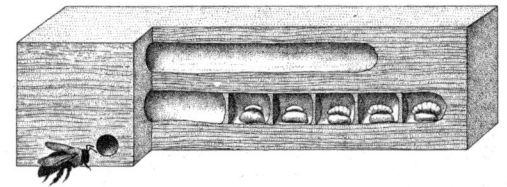

Carpenter bee nest in wood

SWEAT BEES

These small bees have black bodies; the thorax and abdomen have yellowish-white setae.

Nests are usually made in clay soil, and many nests may occur together. Females excavate a long burrow in the ground; along the sides of the burrow are short branches that lead to a brood cell. These cells contain pollen and nectar and a single egg. Several females may use one burrow.

Habits. Sweat bees are attracted to perspiring (sweating) individuals. Although they are non-threatening, they will sting when there is activity near the nest.

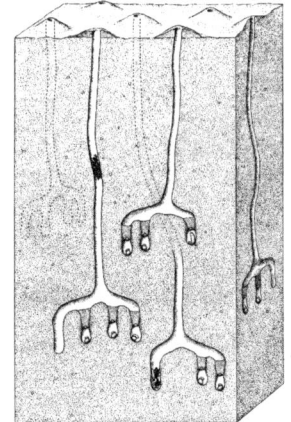

Sweat bee nests

MASON BEES

Body is metallic blue or green; these are small bees with fuzzy bodies.

Nests are small earthen cells in brick veneer of buildings, beetle exit holes in wood, or soil. Females burrow into the soft mortar of stonework and brickwork and line their nest cells with soil.

Habits. These bees have strong mandibles and can tunnel into old or soft mortar. When they excavate exit holes of wood-infesting beetles, frass is scattered and may give the impression of beetle activity.

BUMBLE BEES
Body is covered with setae, usually black with yellow bands.

Nests are usually in old mouse burrows in soil. Colonies last one season; at the end of summer, colonies contain about 100 workers.

Habits. Nests often are built close to buildings and in soil around ornamental shrubs.

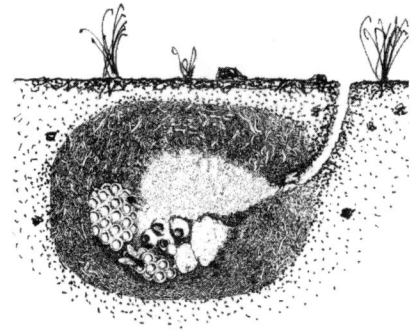

WASPS

MUD DAUBERS
Body is shiny black or dark blue, or it has black and yellow markings.

Nest cells are provisioned with spiders. After the female constructs the chambers, she captures and paralyzes spiders to place in the cells as a larval food. An egg is laid in each cell, and it is sealed.

Habits. These wasps build mud cells in parallel rows, some species build small rounded nests made of mud. These wasps capture spiders for their nests, including black widow spiders.

CICADA KILLER WASP
These are large wasps. The body is dark brown with reddish-orange legs, and yellow markings on the thorax and abdomen. There are several species of cicada killer wasps, with one species occurring east of the Rocky Mountains, two in the western U.S., and two in Florida.

Nests are burrows in the soil. Nesting occurs in late summer and fall, at the time that annual cicadas emerge and begin singing in trees. The prey is cicadas captured in trees.

Habits. The female flies in a circle around the tree or shrub. When a cicada is located, the wasp stings it, then grasps it to fly it back to the burrow. Females may dig burrows that contain 15 cells provisioned with one to three cicadas each to feed the young. Females are generally not aggressive, but males will protect areas around burrows.

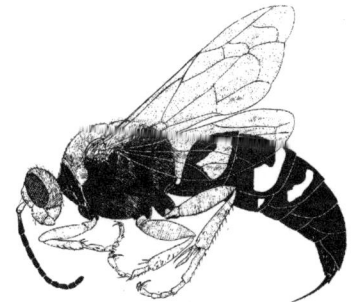

UMBRELLA WASPS
Workers are brown to blackish brown with yellow markings on the abdomen.

Nests are originated in spring by one or more queens. Cloudy and rainy weather at this time will limit the number of new nests formed, and queens may share a nest. These nests develop quickly and often are large.

Habits. Nests are used one season and queens overwinter. A new nest is often built next to the queen's original nest, and the result may be a large number of new and used nests in one site. This wasp preys on caterpillars and also feeds on honeydew and bruised fruits. Queens from nests along house soffits often will overwinter indoors and become active in late fall and spring.

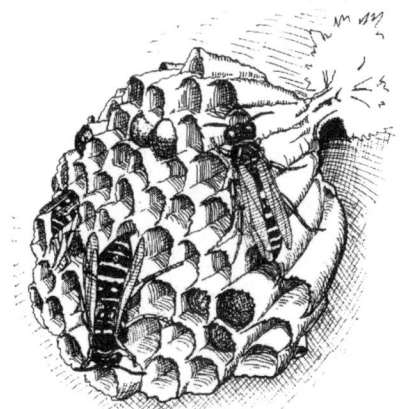

Bumble bee nest

Mud dauber nest

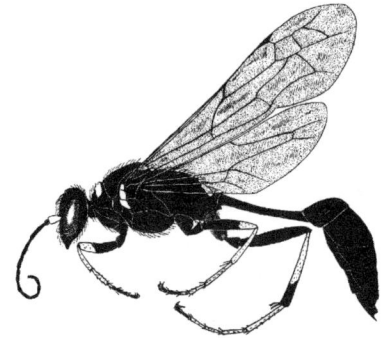

Mud dauber wasp

Cicada killer wasp

Umbrella wasp nest

37

EUROPEAN HORNET

Body is brownish black with narrow to broad yellow bands on the abdomen.

Nests usually are built in trees, attics, and house walls. A brown "envelope" covering the nest distinguishes this species from large yellowjackets. A mature nest has about 1,000 workers. Nests often have a foul odor, and when they are built in attics, they also can extend into living spaces.

Habits. Prey includes grasshoppers, cicadas, flies, yellowjackets, and honey bees. Workers fly and hunt for prey at night, and they are attracted to outdoor lights and lighted windows. These are very large hornets, and their behavior of hunting at night and moving to lights can be threatening.

European hornet distrihbution

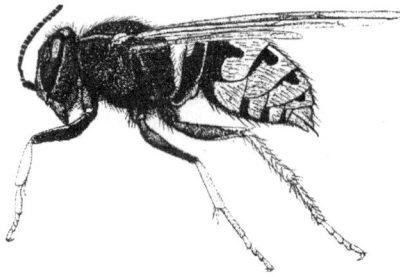

European hornet and abdomen

BALDFACED HORNET (YELLOWJACKET)

Workers are black with white markings. The dorsal surface of the first to third abdominal segments is entirely black.

Nests usually are established in ornamental bushes and trees, electric power poles, houses, sheds, and other structures. Colonies decline in September. The majority of nests contain fewer than 2,000 cells.

Habits. Workers forage for flies and other yellowjackets.

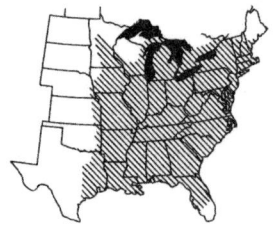

Baldfaced hornet distribution west and east

Baldfaced hornet and abdomen

AERIAL YELLOWJACKET

Body is black with yellow markings. The yellow bands on the abdomen segments are distinct.

Nests may be in shrubs or bushes, the tops of trees, and structures. Mature nests can have 644 to 4,290 cells, and there may be as many as seven combs.

Habits. The large paper nests of this species usually are built well above the ground and away from the reach of potential predators. Workers forage during the day for grasshoppers, tree crickets, caterpillars, flies, and spiders. They are attracted to sugar in late summer. By late summer, there are no larvae in the nest, and workers no longer forage for protein to feed young, so more carbohydrates are eaten simply as an energy food.

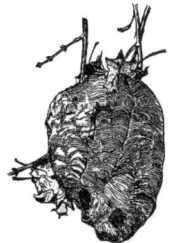

Aerial wasp nest

Aerial yellowjacket and abdomen

GERMAN YELLOWJACKET

Body is brown with yellow markings on head and thorax, and bands on the abdomen. The markings on the abdomen are distinct for this species.

Nests may contain as many as 2,000 workers. They are often built in attics and wall cavities. Colonies usually last one year.

Habits. This species preys on a variety of arthropods. Most workers forage close to the nest site.

German yellowjacket distribution

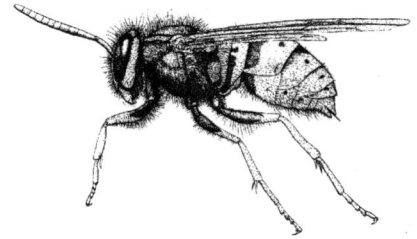

COMMON YELLOWJACKET

Worker abdomen has distinct yellow bands; the first segment of each antenna is entirely black. The thorax lacks the two stripes that are characteristic of the Southern yellowjacket.

Nests usually are built in decaying logs, stumps, or downed trees, and, less commonly, in walls or trees. The nest envelope usually is made of rotten wood fiber and is brittle. Nest populations can be nearly 3,000. Colonies are founded in May or June and peak in September; some are active until October.

Common Yellowjacket thorax

Habits. Prey includes caterpillars, flies, and other flying insects. As the name indicates, this is one of the most common species.

WESTERN YELLOWJACKET

Body is black and yellow. The first antennal segment is yellow at the end.

Nests usually are built in subterranean cavities, but occasionally in walls of buildings. Nests may contain nearly 4,000 workers when the colony is at its peak.

Western yellowjacket distribution

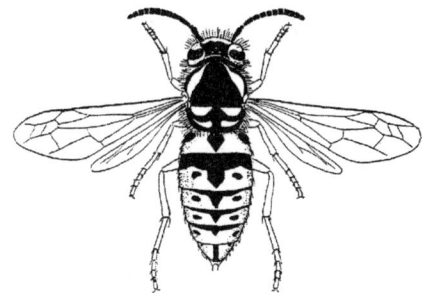

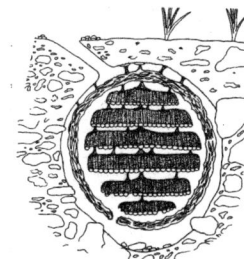

Habits. The workers of this yellowjacket prey on spiders, grasshoppers, and flies. Periodic outbreaks occur every three to five years, following a warm and relatively dry period in spring. Cloudy, wet weather in spring limits the food available for founding queens, and the number of successful nests decreases. Following a relatively warm, dry spring, there may be an increase in the number of successful colonies, and in late summer and fall, the workers can be a nuisance around buildings and recreation areas.

German yellowjacket and abdomen

Common yellowjacket and abdomen

Western yellowjacket and ground nest

EASTERN YELLOWJACKET

There is an anchor-shaped black mark on this yellowjacket's first abdominal segment. The first antennal segment is black.

Nests usually are subterranean. In urban environments, nests are in the walls of houses and other buildings. Colonies usually have peak numbers of workers in August or September. Nests contain 2,000 to 5,000 workers.

Habits. This species often builds nests underground around the perimeter of houses and commercial buildings. The opening to the nest may be hidden among flowers, often in a shaded location or protected from rainfall. In these sites, a nest may be disturbed accidentally and the wasps become aggressive. Colonies contain a large number of workers in fall, and they remain active until the first or second frost.

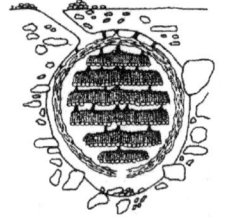

Ground nest - yellowjacket

SOUTHERN YELLOWJACKET

Body has yellow and black markings; the abdomen has narrow yellow bands. The thorax has two distinct parallel stripes.

Nests are located in turfgrass surrounding houses and buildings, recreation areas, and roadsides. Colonies contain 500 to 4,000 workers.

Habits. This yellowjacket is a social parasite of other yellowjacket species. The parasite queen takes over the nest from the host queen and assumes complete control of the colony. This species will vigorously defend its nest, especially in fall when there are many workers in the colony.

Southern yellowjacket thorax

Southern yellowjacket distribution

Eastern yellowjacket and abdomen

Southern yellowjacket and abdomen

CHAPTER 4

BED BUGS, STINK BUGS, AND BOXELDER BUGS

Bed bugs are blood-sucking parasites of humans and other animals. The nymph stages and adults all require blood to develop and survive. They live in cracks and crevices near their hosts and travel at night to get blood meals. These small insects have well-developed eyes, and their antennae can detect the heat and carbon dioxide given off by animals. Bed bugs depend on humans to survive and spread; there are no natural or reservoir populations of this insect. They are spread in luggage, bedroom furniture, and bed frames. Adults live nine to 12 months and can survive long periods without feeding.

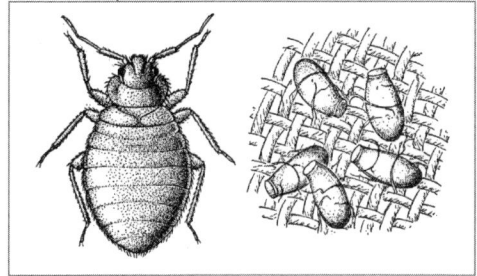

Bed bug and eggs

Stink bugs suck the sap from ornamental plants and agricultural crops. The brown marmorated stink bug is recognized by pale bands on the antennae and abdomen. It develops from egg to adult during summer. Adults are formed in fall and, at that time, they look for a protected place to spend the winter. They gather in large numbers on the sides of building, and then move indoors around doors and windows. These bugs become active and crawl around during warm periods in winter.

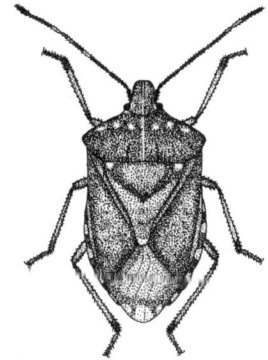

Boxelder bugs are red and black with red lines on the back; immature stages have red abdomens. They suck the sap of leaves and seeds of various trees, including maple and boxelder trees. Adults and nymphs are found primarily on the trees that have seed pods. In fall, the adults look for a place to spend the winter; they gather in large aggregations on the sunny side of buildings, then may move inside through cracks and crevices around windows and doors. On warm winter days and in early spring, they become active and may crawl about the house.

Stink bug adult

Kudzu bugs are closely related to brown stink bugs; they have similar feeding and overwintering habits. These bugs suck the sap of the kudzu plant, which grows as a vine in many regions of southeastern U.S. However, they can feed on other plants. When adults search for overwintering sites, they often come to the perimeter and sides of houses and large buildings.

Backswimmers and water striders occur naturally in ponds and lakes, but they also can invade swimming pools. They can fly to pools from nearby natural habitats. Backswimmers are predators of other insects, but they can bite people.

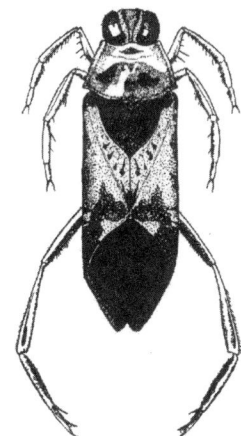

Backswimmer

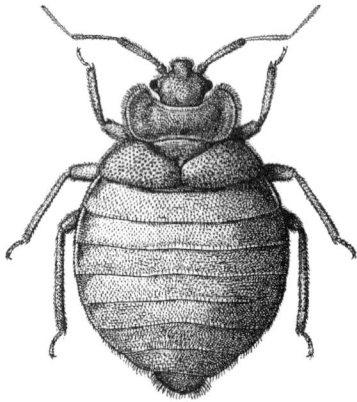

Bed bug

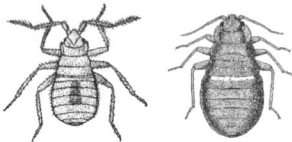

Bed bug small & large nymph

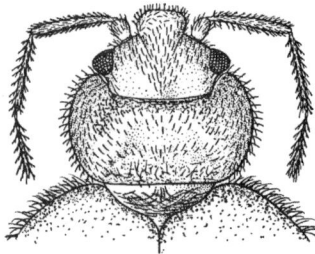

Tropical bed bug head

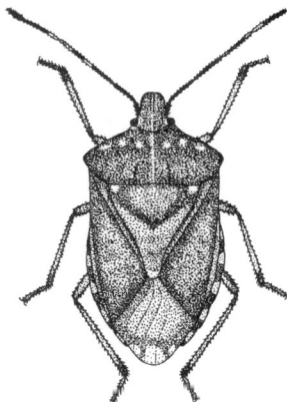

Brown stink bug adult

Brown stink bug nymph

BED BUG

Adults are oval and somewhat flattened. Wings are absent, but they have small wingpads. The body is reddish brown to dark brown depending on whether or not it has taken a blood meal. Nymphs are similar in shape to the adults.

Eggs are yellowish white, with a distinct cap at one end. Females deposit eggs in batches of 10 to 50; they can lay about 350 eggs, but the range is 200 to 500. Hatching is in six to 17 days. Development to adult is through five nymphal instars, and takes about 14 days.

Development is not completed at temperatures below 55°F. Adult males can live for about 176 days and females about 277 days without feeding. There are three or four generations per year.

Food (blood) is required between each molt and before egg development. Between molts, nymphs feed about every six days. Feeding is nocturnal and usually peaks soon after sunset and before dawn, but bed bugs will feed during the day.

Habits. In cool weather, the adults and nymphs may remain active in the harborage for several weeks. Carbon dioxide and warmth are detected and used to locate a host. These bugs respond to temperature gradients only two degrees above normal. Adults produce an aggregation pheromone and an alarm pheromone. Bed bugs prefer harborages with rough surfaces, such as bed frames and mattresses and along baseboards close to the bed.

Tropical Bed Bug (*Cimex hemipterus*) is distributed primarily in the southern hemisphere, but is known to occur in southern U.S. It is rarely associated with bed bug infestations in commercial and residential buildings. This species can be distinguished from the common bed bug (*Cimex lectularis*) by the sides of the pronotum. In the tropical bed bug, the sides of the pronotum are not expanded, but in the common bed bug, they are large and spread out. The setae on the pronotum are straight and sparse.

BROWN MARMORATED STINK BUG

Adult body color is an irregular pattern (marmorated) of dark and pale brown. Nymphs are grayish black with a white spot on the apical third of the antenna.

Eggs are barrel shaped, yellowish red and deposited on the underside of leaves in batches of about 25. Hatching occurs in about five days. Egg laying begins in June and continues to September. A female can produce about 400 eggs in her lifetime. There are five nymph stages.

Food includes seed- and fruit-bearing plants. Stink bugs have piercing-sucking mouthparts and feed by sucking the sap from plants.

Habits. Adults overwinter in large numbers inside and outside of buildings. Overwintering flights of adults to new or previously infested sites begin in October and lasts for about three weeks. Movement to overwintering sites begins during the day when the temperature reaches about 77°F.

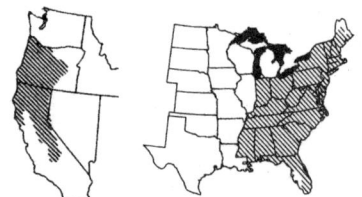

Brown stink bug distribution west & east

APHIDS

These are soft-bodied insects that are green, brown, black, and sometimes pale yellow to white. They have piercing-sucking mouthparts and can pierce the surface of plants and almost continuously suck plant sap.

Most aphids produce droplets of honeydew as they feed. Ants use the honeydew droplet produced by aphids as food. Aphid populations rely on ants to remove the honeydew; ants rely on honeydew as a food source. This is a food for many species of ants that are indoor pests, including pavement ants, carpenter ants, odorous house ants, ghost ants, and white-footed ants.

Aphid group on stem

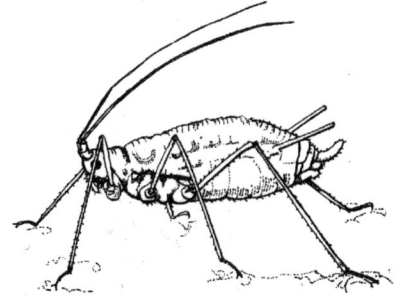

Aphid

BOXELDER BUG

Adult body is gray-brown to black with three red lines on the pronotum and forewing; the abdomen is usually red. The small nymphs are bright red and become marked with black when about half grown.

Eggs are laid on the bark and leaves of the tree; total eggs per female are 200 to 300. Hatching occurs in 11 to 14 days. Development from egg to adult takes about 60 days, and there can be three generations per year. Adults and large nymphs of the last generation seek an overwintering site in late fall.

Food includes the seeds, leaves, and twigs of boxelder and other maple trees, including silver maple, sycamore maple, and ash. This bug also will feed on young fruits such as apple, pear, peach, plum, and grape. Adults of the first generation feed on fallen boxelder seeds on the ground or on low vegetation. They feed on female boxelder and other maple trees once the seeds begin to form.

Habits. Overwintering occurs in large numbers around building foundations and ground-floor windows. Boxelder bugs gather on the south and west sides of buildings where the sun heats exposed surfaces; they are sensitive to small temperature differences and select the warmest substrate. Adults are capable of flying about two miles to find a suitable overwintering site.

Winter harborage. Overwintering sites selected by the boxelder bug, kudzu bug, marmorated stink bug, and cluster fly usually are on east- or west-facing walls of buildings or attics. Overwintering insects can detect small differences in surface temperature, and the radiant energy from the sun reaches these sides during half a day. Thus, these sides provide the greatest warmth in winter and quickly warm up in spring. Horizontal or flat surfaces are warmest during the middle of the day.

Boxelder tree leaf

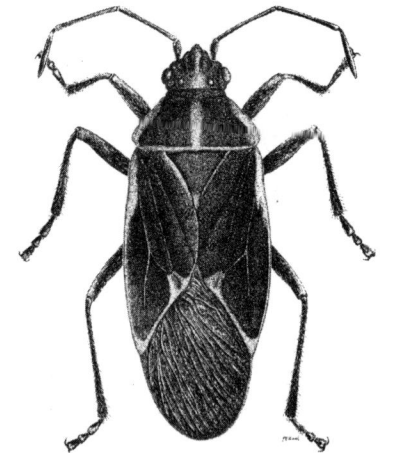

Boxelder bug

Boxelder bug distribution

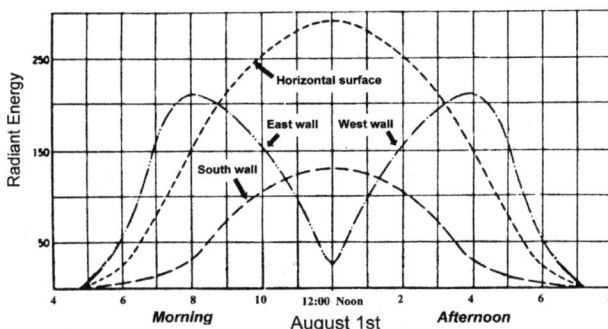

Overwintering harborage

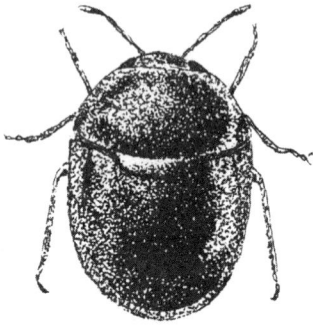

Kudzu bug

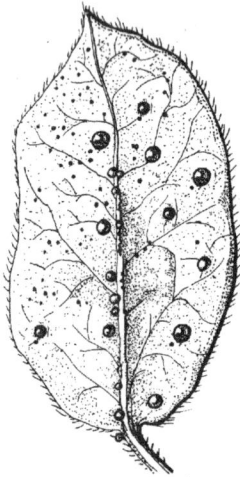

Spots on leaves

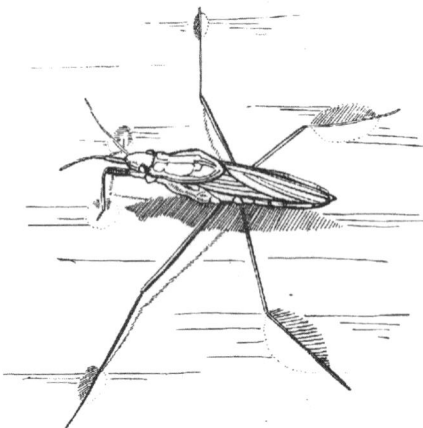

Water strider

KUDZU BUG
Adult body color is an irregular pattern of green and brown with a rounded square shape; they are about 1/4 inch long. They move slowly and fly when disturbed.

Eggs are deposited on the undersides of leaves. Hatching occurs in about five days. Development from egg to adult takes about six weeks.

Food includes kudzu, wisteria, and related plants. The first generation of kudzu bugs feed on kudzu or wisteria; the second generation may feed on soybeans in addition to kudzu.

Habits. Overwintering flights of adults to the sides of buildings begins in October and lasts for several weeks. They become active in early spring and move to food plants.

This bug produces an odor when disturbed. When crushed it will stain surfaces; the body fluids also may cause skin irritation in some individuals. The adults fly to light-colored surfaces, especially white clothing or the sides of houses. Adults overwinter in large numbers inside and outside of buildings.

Distribution of this bug is currently limited to southeastern U.S. It is expected to spread to regions where the kudzu vine grows, and this includes the Northeast and Midwest.

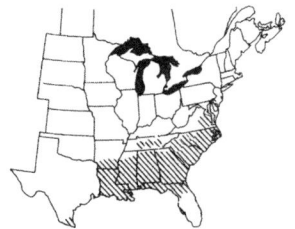

Kudzu bug distribution

SPOTS ON LEAVES
The leaves of flowering plants, shrubs, and ornamental trees may have yellow and brown spots—dead areas along the veins and the perimeter. These are usually caused by the feeding of insects that have piercing-sucking mouthparts. Their mouthparts are inserted into the leaf surface and plant sap is sucked out; brown or yellow spots usually indicate feeding.

Spots on leaves often are blamed on insecticide spraying around ornamental plants. However, modern insecticides and solvent systems are not phytotoxic and do not cause spots on leaves.

WATER STRIDER
Adult is about 3/4 inch long, and brown to blackish brown. The mid and hind legs are long; the front legs are short. The adults have short wings, but they can fly.

Habits. These insects are capable of walking on the surface of the water; they occur in lakes, pods, and slow moving streams. They are predators of other insects. Water striders are sometimes found in swimming pools, but they do not bite.

BACKSWIMMER

Adult is about 1/2 inch long; the body color is black to brown and sometimes shiny blue. The hind legs are long and adapted for swimming.

Habits. These bugs live in water as predators; they are fast swimmers and swim upside down. They can deliver a painful bite to people they contact in water. They are capable of long-distance flight. Backswimmers commonly occur in backyard and commercial swimming pools, even those that are a long distance from a pond or lake.

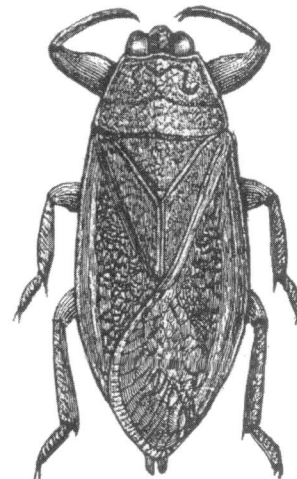

GIANT WATER BUG

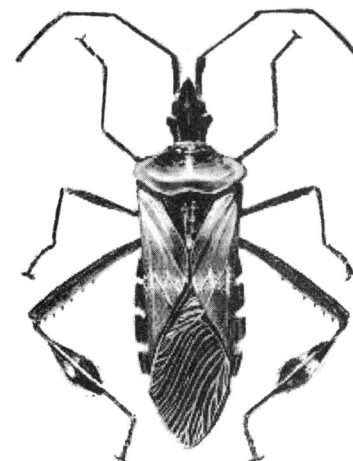

Body is uniformly brown, and about three inches long. The front legs are enlarged at the base, and the wings are large and overlap the abdomen.

Habits. These insects inhabit lakes and ponds where they are predators of other insects and small animals, including frogs. They have piercing mouthparts and suck the blood from their prey. They can fly long distances and can be found far from water. These bugs are attracted to lights at night, especially commercial lighting; they fly around lights making a loud buzzing noise.

WESTERN CONIFER SEED BUG

Adult body color is brown, the abdomen is banded and is seen at the sides of the wings. The antennae are banded with white. The hind legs have expanded sections with white bands.

Eggs are laid on the needles and cones of evergreen trees, including spruce, pine, and fir, but pine trees are the preferred site. Eggs hatch in ten days; there are five nymph stages that look like the adult but lack wings.

Food includes seeds of pine cones and other conifers. The nymphs and adults have sucking mouthparts and suck plant sap from the seeds.

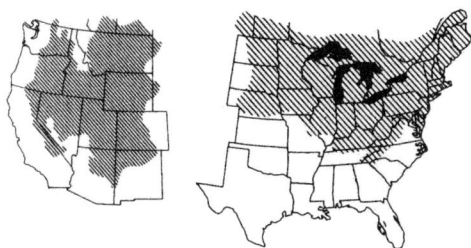

Western conifer seed bug distribution west & east

Habits. Adults overwinter in bird nests and rodent burrows and under bark of logs and firewood. They often move indoors around doors and windows. The adults make a loud buzzing noise when they fly. They do not bite or sting.

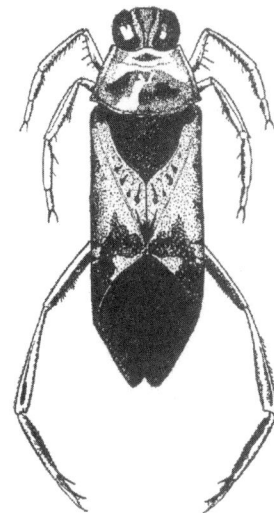

Backswimmer

Giant waterbug

Western conifer seed bug

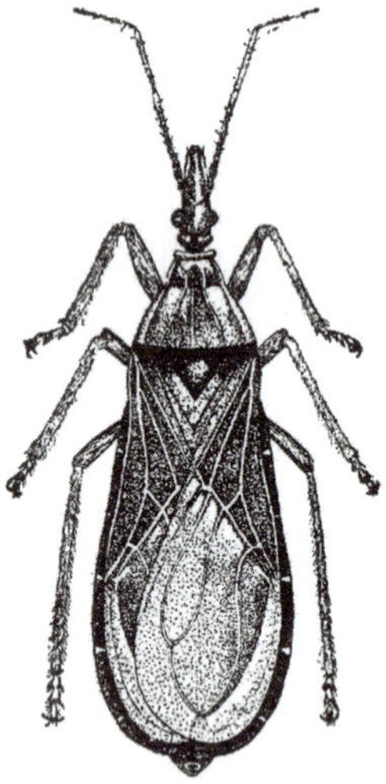

Kissing bug

KISSING BUGS

The adult is about one inch long with an elongated head and beak. The body is dark brown to black, the abdomen is wide and has flattened sides sticking out beyond the margins of the wings.

Eggs are laid in cracks and crevices; hatching occurs in eight to 30 days; nymphs develop though five instars. Development can require up to three years to complete.

Food. The name "kissing bugs" comes from their habit of biting and taking blood around the mouth. They feed on the blood of humans and other animals.

Habits. These blood-sucking insects can transmit Chagas' disease (*Trypanosoma cruzi*) to small animals, dogs, and man. Adults are attracted to lights at night. Feeding occurs at night when the host is sleeping; the adults and nymphs are able to detect heat and carbon dioxide from their hosts.

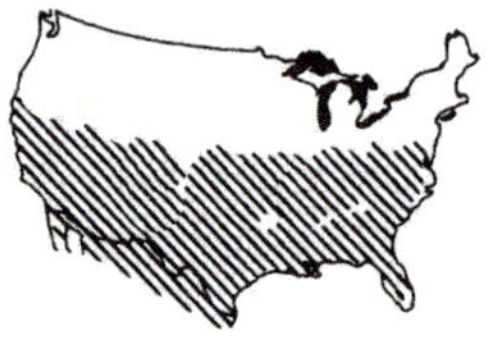

Kissing bug distribution

CHAPTER 5

BEETLES

Beetles are the largest group of insects, there are over a quarter million species. They occur almost everywhere: some live and breed in lakes and ponds; some feed on the leaves of trees and shrubs; some feed in trees, firewood, and structural wood; and some feed on grain, flour, wool, and leather. The most common indoor beetle pests are those that infest stored food and natural fabric.

Beetles have four stages of development: egg, larva, pupa, and adult. There can be several generations a year, especially when they live indoors. Adults and larvae have chewing mouthparts, but adults usually do not feed much and do not live long. The larval stages cause damage by their feeding, and these stages can live for months or even years, as in the case of some wood-infesting beetles.

Wood is a source of food for several beetle species. Some attack the trunk of live trees, others infest logs, and other species infest processed wood used for house framing, furniture, and flooring. In all these cases, it is the feeding tunnels of the larval stage that decreases the strength of the wood. **Wood-infesting species**, such as powderpost beetles, can complete several generations in wood framing and threaten the stability of the structure. The larvae have the ability to digest cellulose, or in the case of lyctid powderpost beetles, they utilize only the starch in the wood.

Beetle-infested wood

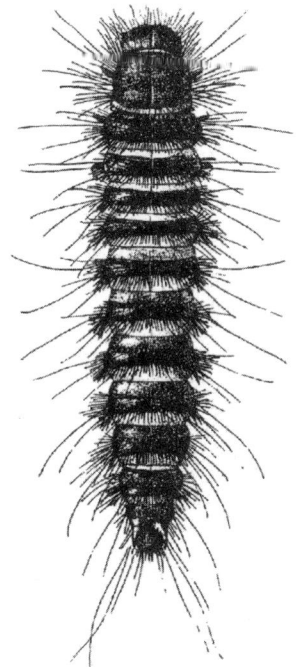

Stored flour, bread, noodles, and other grain products can be infested with beetles. There are several species of **flour beetle** that are common in households around the world. It is almost impossible to store or process flour without having some level of infestation. Although it is the larvae that feed on the product, it is the adult beetle that penetrates storage bins and packaging materials to spread an infestation. Whole grains, such as wheat and rice, can also be attacked by beetle adults and larvae.

Wool fabric, leather, and other natural materials are attacked by several species of **dermestid beetle**. Many of these species are called carpet beetles because of their association with carpeting that was at one time predominately made of wool. The carpet beetle is one of only a few insects that can eat and digest wool, leather, and animal hair. However, it is not limited to these materials, but will feed on a variety of stored food products, dead insects, and dead animals that may occur indoors or outdoors.

Larder beetle larva

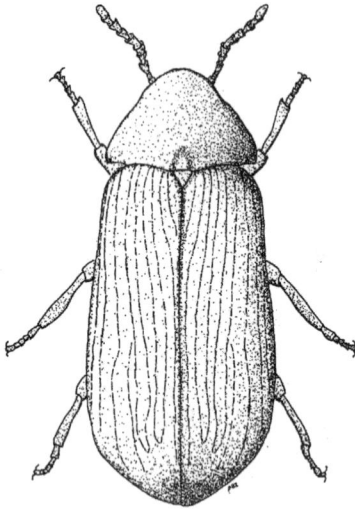

Anobiid beetle

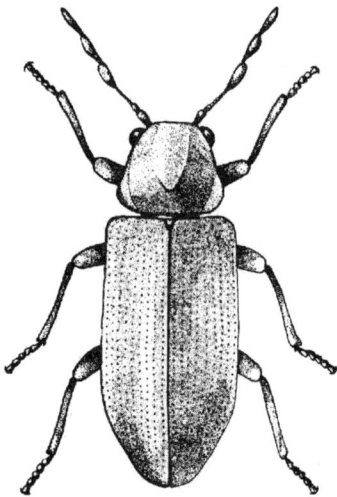

Furniture beetle

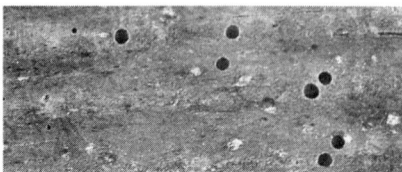

Anobiid beetle wood damage

Anobiid beetle larva

ANOBIID BEETLES

Anobiid (family Anobiidae) infestations may be found in the inspection of joists and rafters in crawlspaces, basements, and attics. Infestations can be recognized by 1/8-inch, round holes which have frass around the outside or accumulating beneath. These beetle exit holes are usually the only sign of damage. The adult beetles live only a short time, but the larvae may live in galleries in the infested wood for up to three years.

Development. Females lay eggs on the wood surface or sometimes inside an old exit hole. The eggs hatch in about two weeks and the first-stage larvae immediately bore into the wood. The larvae feed for one to three years, depending on the wood moisture and age of the hardwood. Full grown larvae tunnel to the wood surface to pupate, then as adults, they cut emergence holes to the outside.

Frass. The frass of anobiid beetles contains coarse, hard pellets, which feel gritty when rubbed lightly between the fingers.

Habits. These beetles infest all types of seasoned wood, both hardwoods (oak, maple) and softwoods (pine, spruce). However, softwood timbers used in house framing are the most often infested. Larvae can feed in wood with moisture content as low as 10%, eggs require about 45% relative humidity to hatch.

FURNITURE BEETLE

Adults are uniformly reddish brown. The wing covers have distinct longitudinal lines, and the last antennal segments are enlarged.

Development. Eggs are deposited on rough wood surfaces or in cracks. Eggs hatch in three to six weeks. Larval development takes about two years, but can extend to five years. Larvae produce galleries filled with frass; they bore close to the wood surface to form a pupal chamber. Adults emerge in May or June; they live about four weeks.

Habits. Infestations are more common in structural pine and spruce framing wood than in furniture. The first-stage larva ingests yeast deposited on the egg by the female; these yeasts become established in the larval gut. Yeasts are killed at 77°F, which restricts the initial development or re-infestations of timbers in some attics.

ANOBIID POWDERPOST BEETLE

Adults are reddish brown to black. The head of the adult is bent downward. Larvae are C-shaped and pale white, with small legs.

Development. Eggs are laid on the rough surface of wood that is two to five years old. Hatching is in about eight days. The first-stage larvae immediately burrow into the wood. The larval galleries are packed with frass. Larvae reduce feeding in response to low temperatures and low wood moisture. Full-grown larvae tunnel close to the surface and prepare a pupal chamber. The life cycle is one to five years, depending on the wood infested.

Habits. Pine timber may be re-infested and infestations continue until nearly all the sapwood portion of the wood has been consumed, leaving only the annual rings.

ANOBIID BARK BEETLE

Adults are reddish brown to black. The head of the adult is bent downward.

Development. Eggs are laid in the bark. Larvae feed in tunnels between the bark and wood surface. Development takes one to two years.

Frass. The frass is a mixture of brown particles from the bark and white particles from the wood below the bark.

Habits. This beetle infests pine, spruce, and fir, especially lumber that has small sections of bark remaining. The damage is often confused with active infestations of powderpost beetles. This beetle does not reinfest the wood and does not cause structural weakening of lumber.

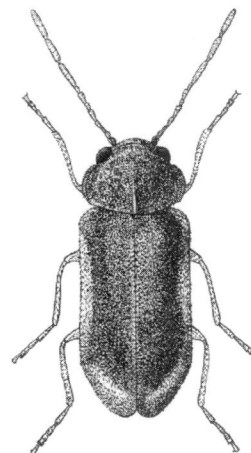

Anobiid bark beetle adult

LYCTID BEETLES

Lyctid (family Lyctidae) beetle infestations are in hardwoods, and usually hardwood flooring. These beetles are common in imported hardwoods used in household molding. Infestations are characterized by 1/32-inch, round holes with frass surrounding the opening (in flooring) or falling from holes. The adult beetles live only a short time and rarely are found.

Development. Females lay eggs in exposed ends of flooring or other wood, usually in the pores of the grain or in cracks and crevices. Hatching is in about three weeks. Development from egg to adult requires six to 12 months, depending on the starch and moisture content of the wood. The life cycle extends to two years, or as much as four years under unfavorable conditions.

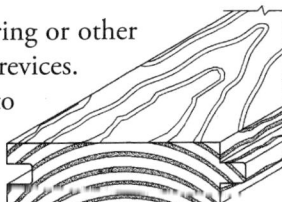

Hardwood flooring

Bark beetle wood damage

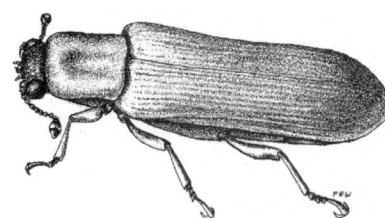

Frass. The frass of lyctid beetles contains no hard pellets and is soft when rubbed lightly between the fingers.

Habits. Females prefer to lay eggs in wood that is less than five years old, because it has adequate starch content (at least 3%) for the developing larvae. Bamboo, which is a grass, has high starch content, and also can be infested by lyctid beetles.

Lyctid beetle

LYCTID POWDERPOST BEETLES

Adults are reddish brown to black; the body is distinctly flattened.

Development. Eggs are laid in cracks on the wood surface; hatching occurs in 14 days. Larval development is eight to 10 months. Full-grown larvae bore close to the wood surface and form a pupal chamber. Development from egg to adult is one year, but can extend to two to four years. Adults emerge from May to September.

Habits. Larvae survive wood moisture content between 8% and 30%; larval development is optimum at 16% moisture content.

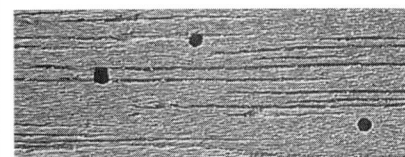

Lyctid beetle wood damage

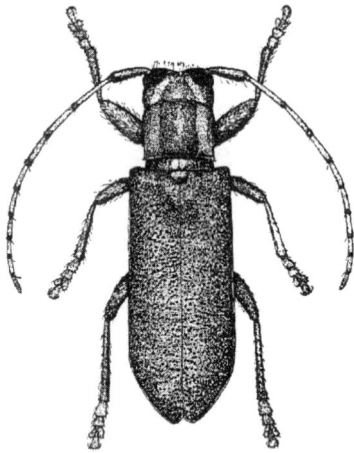

Longhorned beetle adult

Longhorned beetle larva

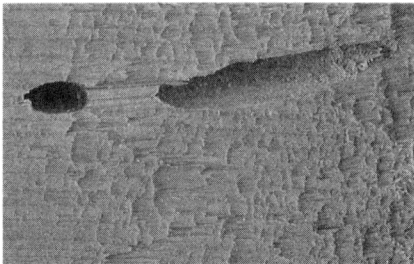

Pine sawyer wood damage

Pine sawyer holes with frass

Old house borer wood damage

LONGHORNED BEETLES

Old house borer and longhorned beetle damage may be found in the inspection of structural wood in crawlspaces, basements, or attics. Emergence holes of these beetles are oval and about 3/8 inch long. They may be empty, be plugged with fibrous pieces of wood, or contain wood powder (frass).

Development. Females lay eggs in crevices in the wood surface. Eggs hatch in about 10 days and the first stage larvae bore immediately into the wood. Larvae feed for one to three years, depending on wood moisture. Full-grown larvae tunnel to the wood surface and cut exit holes.

Frass. The frass of most longhorned beetles is fibrous; the frass of the old house borer is powdery.

Habits. Most longhorned beetles attack dead and downed trees and logs. They have a one- or two-year life cycle and do not reinfest the wood. Lumber made from wood previously infested by pine sawyers will contain the oval galleries made by the larvae. The old house borer infests only structural wood; it does not occur in the forest in dead trees or logs.

PINE SAWYERS

Adults are about one inch long or longer; the antennae are sometimes longer than the body.

Development. Eggs are laid in crevices in the bark. Early-stage larvae feed beneath the bark, late-stage larvae tunnel deep into the wood. Larval development is completed in one or two years. Adults emerge in April and May, and are active throughout the warm season.

Frass. The galleries are usually empty, but fibrous frass may be found in some areas. Frass may be exposed at the wood surface.

Habits. These beetles develop in freshly cut, recently felled, dying, or recently dead trees. Larvae are called sawyers because of the sawing-wood sound they make while feeding. These beetles do not re-infest the original wood.

OLD HOUSE BORER

Adult body is black to brownish black; antennae are not longer than the body.

Old house borer larva

Development. Eggs are placed in cracks and crevices. Hatching occurs in about nine days, and first-stage larvae immediately bore into the wood. Larval development takes two to 10 years. Full-grown larvae tunnel to the wood surface and cut oval exit holes. Adults live about 15 days.

Habits. Larvae feed on wood with moisture content of 10% to 20%. They do not feed on decayed wood. Infestations occur primarily in houses newer than 10 years old.

BOSTRICHID BEETLES

Bostrichid (family Bostrichidae) beetles are primarily pests in hardwoods, but some species attack softwoods. Infestations often are in rough-cut lumber used for pallets, or in hardwood used before it is seasoned. Bostrichids do not reinfest wood after it is seasoned and the moisture content reduced. Adult beetles are 1/4 inch long and cylindrical. There is a patch of short spines on the region above the head, and there may be spines or hooks at the end of the body. Entry and emergence holes are about 1/8 inch in diameter. The larval galleries are filled with powdery frass.

Development. Females bore into the wood surface and prepare tunnels for egg laying. Adult beetles feed on wood as they tunnel. Eggs are laid inside the tunnels, and the larvae create their own tunnels and frass as they feed. Larvae complete development in about 12 months.

Frass. The frass is fine powder, but it does not fall from the entry or emergence holes.

Habits. These beetles infest unseasoned hardwoods, and are not common in household materials. The bamboo borer infests ornamental pieces. Bamboo is a grass but has high starch content and can provide adequate food and moisture for Bostrichid larvae.

BAMBOO BORER

Adults are about 1/8 inch long and dark brown; the front of the thorax has numerous indentations and short spines; the head is bent downward. Larvae are pale white with a dark head.

Development. Eggs are laid in cracks and crevices in bamboo; hatching is in about six days. The larval period is about 41 days; development from egg to adult is completed in 31 to 85 days; adults live 28 to 115 days.

Habits. Infestations can occur in material made of bamboo, including furniture; stored food, flour, and spices also can be infested, but this is rare. The holes in bamboo may be the only evidence, as the beetles are rarely seen.

ORIENTAL WOOD BORER

Adults are about 3/8 inch long and dark brown; the front of the thorax has numerous indentations and spines, and there are two prominent spines projecting forward. The males have hooks or spines at the end of the wings. The head is bent downward, and the antennae are hidden.

Development. Eggs are laid in cracks and crevices and on rough surfaces of wood; hatching is in about seven days. Larval development period is about six months. The adults chew exit holes in the wood surface. Males crawl on the outside of the wood, but females usually remain in the holes to mate and lay eggs.

Habits. This Bostrichid is widely distributed in the Asia-Pacific region and can be found infesting shipments of wood materials from there. Furniture from the Asia-Pacific region may be infested. The damage can remain hidden, but the adults (males) are often found crawling on the wood or close by. This beetle has become established in southern Florida.

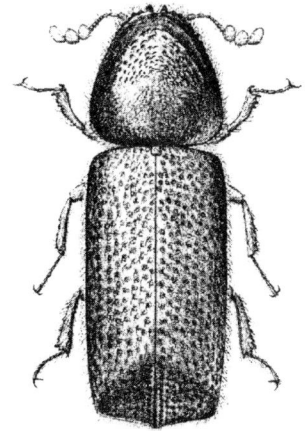

Bostrichid beetle

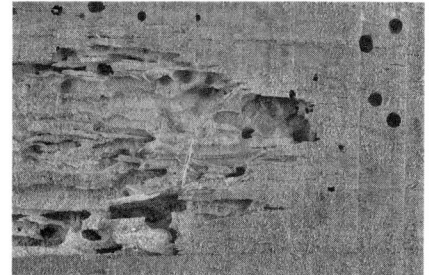

Bostrichid beetle wood damage

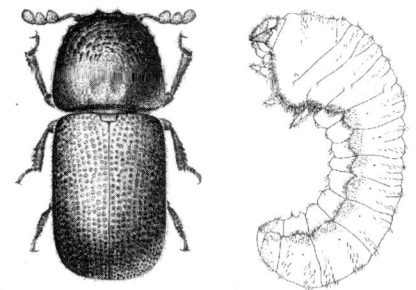

Bamboo borer

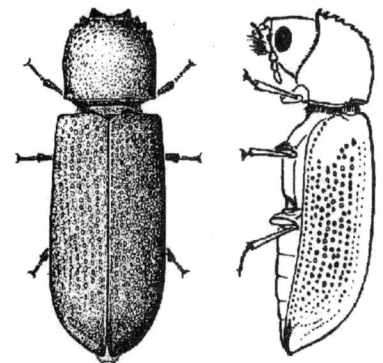

Oriental wood borer

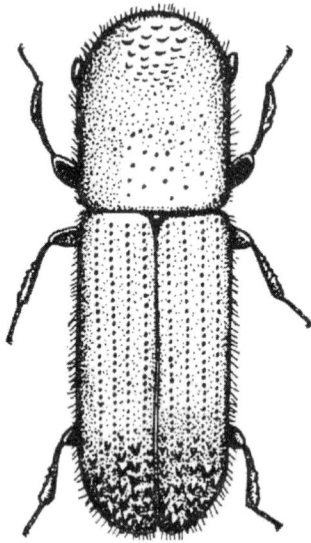

Ambrosia beetle adult

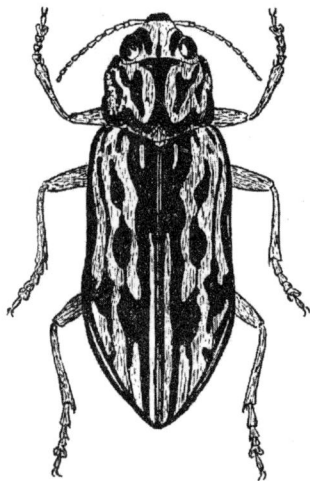

Buprestid beetle

Buprestid beetle larva

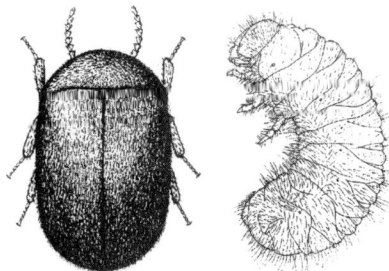

Cigarette beetle and larva

AMBROSIA BEETLES

Adults are dark brown to black. The head is bent downward, and the antennae are hidden.

Development. The female chews a gallery into the outer layer of wood in the host tree. Eggs are laid in secondary tunnels at right angles to the main gallery. Larvae feed on fungi growing on the walls of the tunnel. Larval development is completed in about 30 days. Adults emerge from the tree through the original entrance hole made by the female.

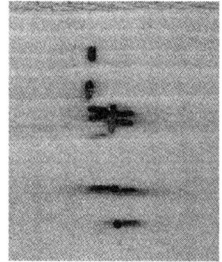

Ambrosia beetle damage

Habits. Damage is done to recently cut trees, including hardwoods and softwoods. When wood is cut for structures, the dark-stained galleries of a previous Ambrosia beetle infestation may be seen. Pine construction timbers often have stained galleries of this beetle.

BUPRESTID BEETLES

Adults are about 3/4 inch long and slightly oval; they are shiny and brightly colored. The larvae are called flat-headed borers because the body region behind the head is broad and flattened.

Development. Eggs are laid in crevices in freshly cut softwood or hardwood logs. First stage larvae bore and feed close to the wood surface; older larvae bore deeper into the wood. Development is completed in one or two years, but it may extend to three or four years in wood with low moisture content.

Habits. These beetles prefer recently felled rather than seasoned wood, and they often come to trees immediately after they are cut. They can occur in seasoned lumber and may be pests of modern log houses. The distinctly oval larval galleries, sometimes packed with frass, may be seen in wood siding. Solitary bees may excavate these exposed galleries as nest sites. The activity of these bees may give the false impression of an active beetle infestation.

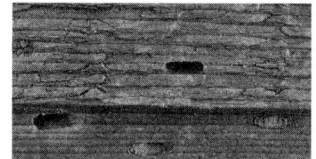

Buprestid beetle wood damage

CIGARETTE BEETLE

Adults are about 1/4 inch and uniformly light brown. The antennal segments are all the same size and shape, which helps to distinguish this species from the drugstore beetle, which is very similar.

Development. Eggs are laid in crevices in food material; hatching is in about 20 days. Larval development ranges from 106 to 135 days. Larvae construct silk chambers for pupation. Adults feed on the same food as the larvae.

Habits. This is a widespread pest of food materials and is not limited to tobacco. First-stage larvae can enter small openings in the seams of food packages. Late-stage larvae are not able to penetrate smooth surfaces or cracks and crevices. Infestations are common in household food materials. The larvae feed on seeds, nuts, beans, spices, yeast, dried insects, fish, vegetables, flour, meal, and tobacco.

DRUGSTORE BEETLE

Adults are about 1/4 inch long and reddish brown; the body shape is somewhat cylindrical. This species is distinguished from the cigarette beetle by its slender shape and the enlarged segments at the end of the antennae.

Development. Eggs are laid singly in crevices of the food material; hatching is in 12 to 37 days. Larval development is completed in about five months, depending on the temperature. A full-grown larva produces a silken cocoon that is covered with particles from the substrate. This makes cocoons difficult to see in infested material. Adults live about 85 days. There are three or four generations per year.

Habits. The small size of the first-stage larvae enables them to enter openings in packaged foods. Larvae move around searching for access to food material; they can survive for about eight days without food. Infestations are common in houses and are also known to occur in medical drugs, grain, spices, tobacco, leather, and textiles.

SAWTOOTHED GRAIN BEETLE

Adult is brown to reddish brown; the pronotum has six teeth on the lateral margins.

Development. Eggs are laid singly or in small clusters; hatching is in about 16 days. Larval development takes 12 days; larvae construct a pupal chamber from food material. Adults live for about 19 weeks. There are six or seven generations per year.

Habits. This is the most common pest of grain and cereal products. It is the beetle that will be found in noodles, cake flour, and meal.

RED-LEGGED HAM BEETLE

Adults are shiny green or greenish blue and have reddish-brown legs.

Development. Eggs are deposited in batches of up to 28 per day. Hatching is in four to six days. Larval development takes 17 to 30 days. Larvae migrate to dry locations to pupate. Adults can live about 14 months. There are two or three generations per year.

Habits. This beetle infests drying meats in long storage or curing meats in a prolonged smoking process.

SPIDER BEETLES

Adults have fine, yellowish-brown setae covering their bodies. They somewhat resemble spiders.

Development. Eggs are laid over a period of four weeks, and females lay 100 to 1,000 eggs. They hatch in about nine days. Larval development requires about 60 days.

Habits. Larvae pupate near the surface of infested material or crawl away to pupate. Adults are most active at night. Infested materials include nuts, beans, cacao, cayenne pepper, chocolate, corn, dried fruit, dried fish, and poultry food.

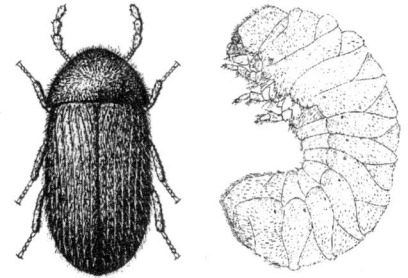
Drugstore beetle and larva

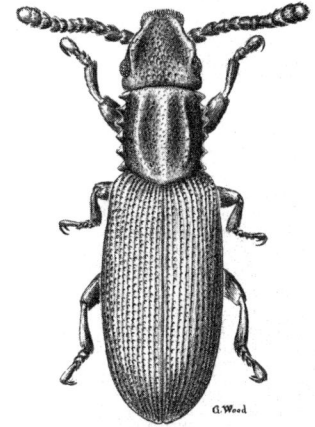
Sawtoothed grain beetle

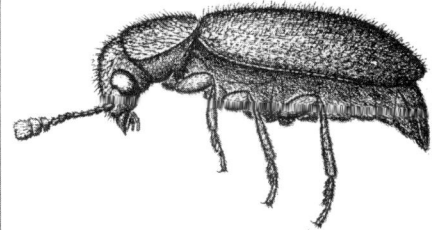
Red-legged ham beetle adult

Red-legged ham beetle larva

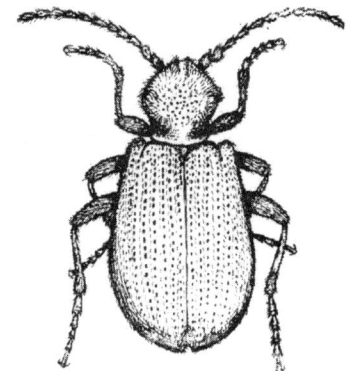
Spider beetle

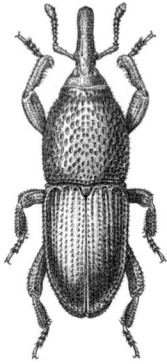

Rice weevil

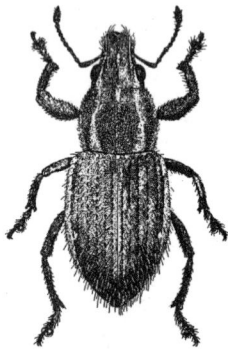

Root weevil

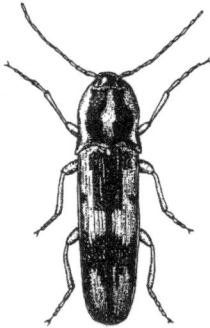

Click beetle

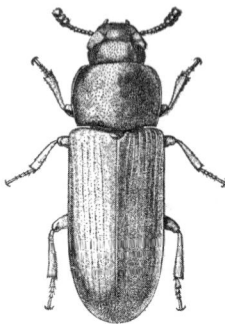

Red flour beetle

Red flour beetle larva

RICE WEEVIL

Adults are reddish brown to blackish brown. Full-grown larvae are yellowish white and C-shaped.

Development. Eggs are laid singly in grain kernels after the female bites a small hole in the surface of the kernel. Hatching is in about six days. No eggs are laid on grain with moisture content below 10%. Larval development takes about 26 days. Adults live four to five months.

Habits. This beetle feeds on beans, nuts, cereals, cereal products, macaroni, cake flour, rice, wheat products, and even fruits, and can infest household cabinets.

ROOT WEEVILS

Adults are dark brown to blackish brown and have a short, wide snout. These beetles do not fly. The larvae feed on plant roots. Adults emerge from May to July, and are present during the remainder of the warm season. When populations peak in spring and fall, adults enter houses and other buildings.

Habits. These short-nosed weevils can be present in large numbers around the perimeter of houses and other buildings, and sometimes move inside. Adults are attracted to lights at night, and often enter buildings through windows and doors.

CLICK BEETLES

Adults are about 1/2 inch long, elongate and brown, sometimes with dark spots. The antennae are long and toothed.

Development. Eggs are laid in soil, and larvae complete development in about four weeks. The larvae then feed on the roots of plants. The adults live for several months but feed very little.

Habits. The adults can bend at the junction of the thorax and abdomen, and then snap back to a straight position, causing an audible snap. Because of this, they are commonly called click beetles. These beetles fly to UV light traps and indoor lights at night; they are often found in houses.

RED FLOUR BEETLE
CONFUSED FLOUR BEETLE

Adults of these two beetles are reddish brown. Each antenna of the red flour beetle has a three-segmented club; the confused flour beetle's has a four-segmented club.

Development. Eggs are laid directly on flour. Larval development is completed in 19 to 22 days. The number of larval instars ranges from five to 11. Pupae are formed in sheltered locations in the infested material; pupal period is four to six days. Adults live for about three years, and females lay eggs for more than a year.

Habits. These beetles infest grain, flour and other cereal products, milk chocolate, dried milk, and hides. Adult red flour beetles can fly short distances, but confused flour beetles have not been observed to fly.

YELLOW MEALWORM

This beetle is somewhat shiny dark brown to black and about 1/2 inch long. The head and antennae are visible. The larva is shiny brownish yellow with a hardened skin; the head is dark brown to black.

Development. Females lay about 300 eggs directly on flour or grain; eggs hatch in about two weeks. Larvae develop through 10 to 14 molts, and they complete development in 10 months. The adults live for about three months; there is one generation per year.

Habits. At night, larvae are active, and the adults fly to lights. Larvae feed on grain, animal feed, and dry pet food. Household infestations are not common, but the beetles are known to feed on food material, particularly dry pet food hidden indoors by mice.

CARPET BEETLES

There are about 50 species of carpet beetles (family Dermestidae) that infest stored food or feed on wool, silk, leather, furs, skins, museum specimens, and carrion. Several species were common pests of household carpeting when it was made primarily of wool fiber. The name carpet beetle is now generally applied to this group of beetles, although their feeding habits extend beyond carpeting. The larvae are capable of digesting keratin, which is the protein in wool and other animal products. The adult beetles fly very well; they visit flowers outdoors to feed on pollen and move indoors to lay eggs on the material that the larvae will infest. Adults of the common carpet beetle are attracted to lights and often are found dead in receptacles of household ceiling lights. The larvae of these beetles often are noticed as they crawl slowly up walls or in kitchen cabinets, dresser drawers, or closet shelves

Development. The life cycle can be extended through six to 20 larval stages. The accumulation of larval cast skins can give the impression of a large infestation, however accumulated cast skins can be from only a few larvae. Larvae are identified by the dense rows of setae that ring the body and the long setae that extend from the sides and the posterior. Setae on larvae are easily detached and can produce allergic reactions, such as rhinitis and respiratory asthma. When there are large infestations, and cast skins of carpet beetles accumulate, setae can become detached and cause irritation.

Habits. Carpet beetle infestations can originate from various organic sources, both indoors and outdoors. In attics and wall voids, these sources include abandoned bird nests, old umbrella wasp and yellowjacket nests, dead mice and birds, and dry pet food cached by a deer mouse infestation. Rat and mouse nests usually contain food scraps and rodent feces, and these provide a food source for carpet beetle larvae. Larder beetle is a species of Dermestid. Larvae of these beetles are not specific in their feeding habits; they can be found attacking animal skins and bee hives remaining in wall voids. In these materials, carpet beetles feed on dead bees and the comb wax. Black carpet beetle infestations often are centered on stored food, and the adults and larvae rarely move away from this site. The varied carpet beetle is responsible for most of the damage to wool, silk, and furs; these larvae are small and easily overlooked. Larvae of the Anthrenus species, such as the common carpet beetle and furniture carpet beetle, can survive nearly one year without feeding. This increases the opportunity for infestations to persist while household material is in storage or transit.

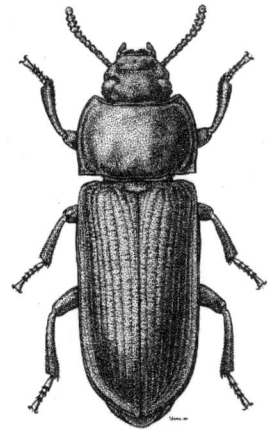

Yellow mealworm adult

Yellow mealworm larva

Carpet beetle

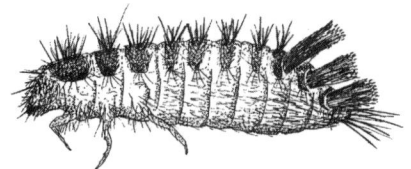

Carpet beetle side view

Carpet beetle side view

Bird nest

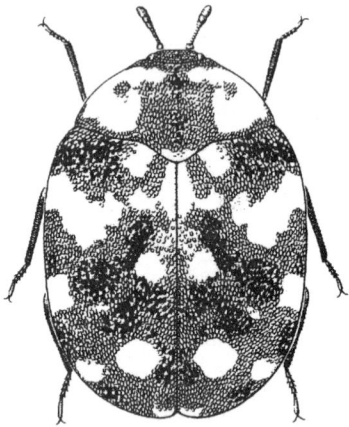

Furniture carpet beetle

FURNITURE CARPET BEETLE

Adult body is rounded oval; there are yellow, white, and black spots scattered on the dorsal surface. Larvae are distinctly banded with a tuft of long setae at the end of the body.

Development. Eggs are laid in batches of up to 57 eggs; hatching occurs in nine to 16 days. Larval development is 112 to 378 days depending on temperature. Development from egg to adult is 93 to 422 days, and adults live 30 to 60 days. Adults emerge in spring.

Habits. Larvae feed in a limited radius and their cast larval skins can accumulate in one place, which gives the appearance of a severe infestation. Larvae feed on wool, silk, fur, feathers, and dry animal material.

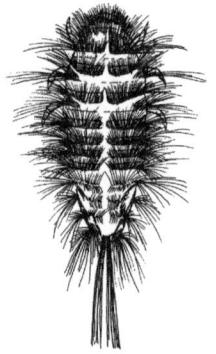

Furniture carpet beetle larva

COMMON CARPET BEETLE

Adult body is oval, gray to black, with a varied pattern of white and orange-red scales on the dorsal surface. Larvae are oval shaped, with long setae at the margins and a tuft of long setae at the posterior end.

Development. Eggs are laid singly or in batches of up to 36. Hatching occurs in 13 to 20 days. The larval development period is through about six instars and takes 60 to 80 days. Development from egg to adult ranges from 89 to 108 days.

Habits. Larvae feed on various animal materials, including wool, feathers, hair, and fur, and museum specimens.

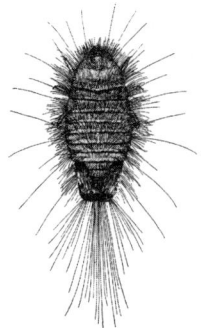

Common carpet beetle

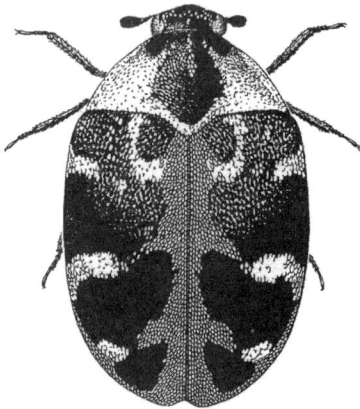

Common carpet beetle larva

VARIED CARPET BEETLE

Adult body has a pattern of white, black, and brownish-yellow spots. Full-grown larvae have a series of light- and dark-brown transverse stripes. At the posterior end of late-stage larvae, there are tufts of setae.

Development. Eggs are deposited on larval food; hatching is in 18 days. Larval development takes 222 to 323 days, and includes five to 16 instars depending on conditions and the food material. Development is completed in one year. Household infestations produce adults in fall.

Habits. This species feeds on stored food materials and grain products, and also on dead insects in light fixtures.

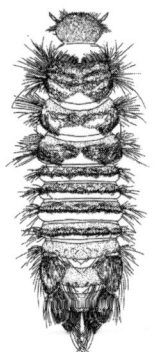

Varied carpet beetle adult

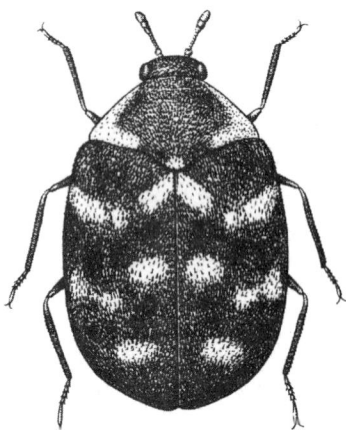

Varied carpet beetle larva

BLACK CARPET BEETLE

Adults are dark brown to black. Larvae are glossy brown and covered with brown setae at the margins; a brush of hairs projects at the end of the abdomen.

Development. Eggs are laid singly directly on the larval food, usually in late fall and winter. Hatching is in six to 22 days. Females lay about 70 eggs. Larval development is 269 to 639 days for males and 258 to 545 days for females. Males live about 40 days, and females live 15 to 38 days.

Habits. Larvae often crawl to the bottom of a food source or other substrate. When disturbed, larvae curl and remain motionless for a long period. Larvae feed on silk cloth, wool, feathers, hair, fur, fishmeal, and cereal products; they often bore into stored food containers.

Black carpet beetle larva

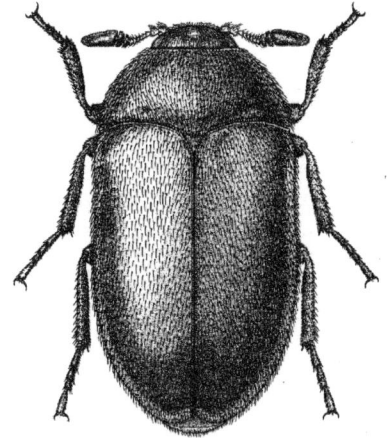

Black carpet beetle adult

LARDER BEETLE

Adult body is dark brown to black, with a pale yellow to yellowish-white band across the back.

Development. Eggs are laid singly on the larval food material; hatching occurs in three to nine days. Development through the pupal stage takes 45 to 50 days. Adults live about 250 days.

Habits. Larvae typically leave the food source to pupate and often tunnel into wood. The larval stages feed on dry meat, fish, cheese, pet food, and hides; indoors, beetles can feed on the bodies of dead mice or birds.

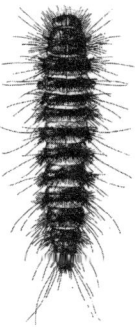

Larder beetle larva

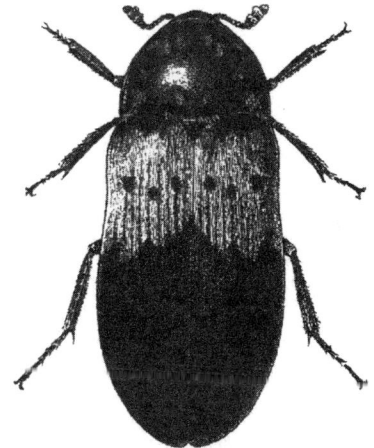

Larder beetle

HIDE BEETLE

Adult body is dark brown to black, and the ventral surface is white. The larva has numerous long setae at the edge of its abdominal segments and on its posterior segment.

Development. Eggs are laid singly or in batches of 17 to 25; hatching occurs in six days. Larval development is about 50 days and there are seven to nine instars. Pupal chambers are constructed after the larvae bore into available surfaces. Males live about 158 days; females live about 180 days.

Habits. This beetle feeds on feathers, fur, bone, cheese, and dried fish. Adults are active flyers and are attracted to natural and ultraviolet light. Larvae of this beetle have been used in museums for cleaning skeletons.

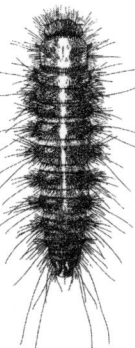

Hide beetle larva

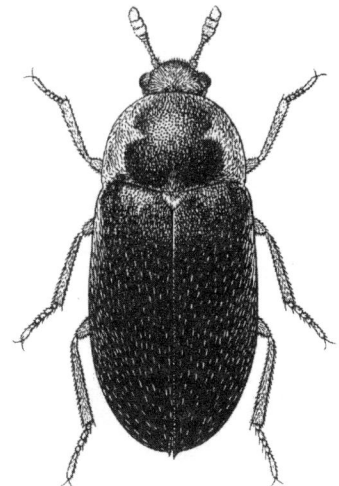

Hide beetle adult

57

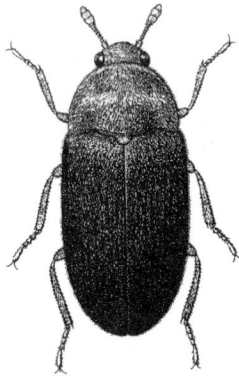

Black larder beetle adult

Black larder larva

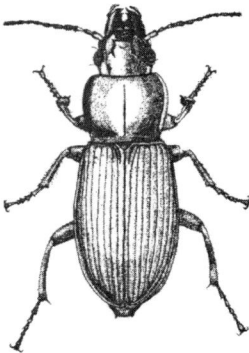

Ground beetle

Elm leaf beetle

Asian ladybird

BLACK LARDER BEETLE

Adults are uniform black or blackish brown, and the underside of the abdomen is brown with brown spots. Full-grown larvae are dark brown and have many long hairs on their back.

Development. Larvae complete development in about 50 days and have seven to nine instars. Males live about 169 days; females live 173 days.

Habits. This is a very common household pest. It feeds on smoked meat, dried fish, bones, hides, animal skins, and cheese.

GROUND BEETLES

Adults are about 3/4 inch long and uniformly black, sometimes shiny. The mouthparts extend forward, and the mandibles are large.

Development. Eggs are laid in soil, and larvae are predaceous on ground-dwelling insects. Larvae complete development in early summer; adults are active for several months.

Habits. Adults are predators of insects, particularly caterpillars. They are usually found in leaf litter, thatch in turfgrass, and mulch. They will move indoors around doors and windows. These beetles are very capable flyers and are attracted to lights at night. They sometimes are mistaken for cockroaches.

ELM LEAF BEETLE

Body is yellowish green to dull green; there is a black stripe along the sides.

Development. Eggs are laid at the base of trees; hatching is in two weeks. Early-stage larvae climb the tree trunk and feed on the underside of the leaves. Full-grown larvae crawl down the trunk to pupate in the bark near the ground or the soil around the tree. Larval development takes two to three weeks; there usually are two generations per year. Second-generation adults select hibernation sites in fall.

Habits. Overwintering locations are leaf litter around the foundations of buildings, and in the attics, eaves, and interior rooms of houses. Adults usually remain active during winter but do not feed.

ASIAN LADYBUG/LADYBIRD BEETLE

Body color pattern varies from yellowish orange to nearly red, and this beetle may have no spots or more than 20. The adult lives several years, and its color may change.

Development. Eggs are deposited in batches of 20 on the underside of tree leaves; hatching occurs in three to five days. Larval development is completed in 12 to 14 days. The pupa is attached to the leaf surface and emerges as an adult in five to six days. There are two or three generations per year; adults live two to three years.

Habits. Adults fly to natural and ultraviolet light indoors. Adult beetles select overwintering sites on the warm sides of buildings.

Borer - Red Oak Trees
do not Kill Tree, Tree already dieying

CHAPTER 6

COCKROACHES

T he only experience most people have with cockroaches is that of German cockroaches in kitchens or American cockroaches crawling out of sewers at night. However, there are about 4,000 cockroach species around the world, and only a few of these are considered to be pests. German and American cockroaches are basically tropical insects. Thus, they select indoor sites for harboring, such as in kitchens and bathrooms that provide the temperature and humidity of their natural habitats.

Adults and the immature stages are brown to blackish brown, oval, and flattened. The head is usually concealed from above by the large pronotum. The antennae are very long, and they have chewing mouthparts. Most adult males have wings and can fly; females often have short wings or none at all and do not fly. Nymphs are similar to adults except for their size and the absence of wings. Eggs are enclosed within an egg case. There are five to 12 nymph stages, depending on species.

Cockroaches locate food by using chemical receptors on their antennae and mouthparts. In general, they prefer carbohydrate foods but also will feed on material high in fat and protein. The availability of food controls their reproduction. The female German cockroach actively forages and eats when she is preparing her egg case, but she remains relatively inactive once the egg case is formed and during the 28 days she carries it. Cockroaches defecate while feeding and moving about, spreading pathogens to surfaces they contact.

Their legs have strong spines and setae, and the pads on the undersides of their feet enable them to climb. A large pad between the claws at the end of each leg helps cockroaches move quickly on smooth and vertical surfaces. This can be seen in the movement of German cockroaches in kitchens and the American cockroaches indoors and out. Oriental cockroaches do not have the large pad between their claws. These cockroaches can easily climb rough horizontal surfaces, but they can not climb smooth vertical surfaces. They often become trapped in sinks and bathtubs that have smooth sides, and give the false impression they have come up the drain pipe.

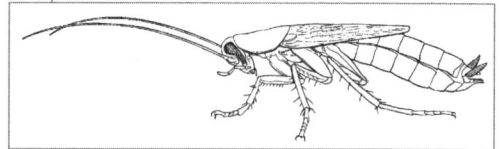

Cockroach body side view

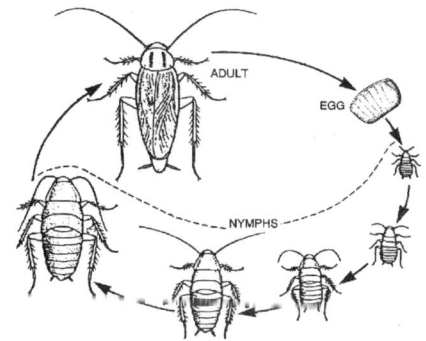

Cockroach development

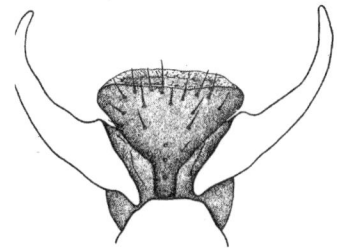

American cockroach claw

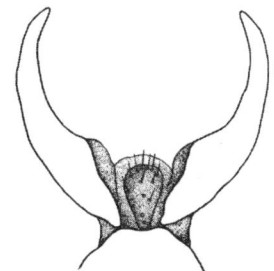

Oriental cockroach claw

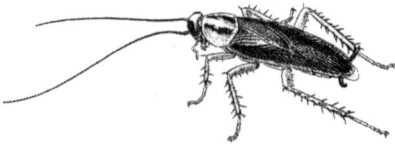

German cockroach

German cockroach nymph

German cockroach with egg case

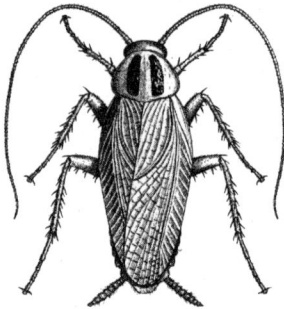

Asian cockroach

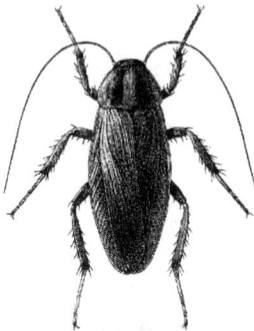

Field cockroach

Field cockroach head

GERMAN COCKROACH

Adults are about 1/2 inch long and light brown to yellowish brown; females are slightly darker than males. Male body shape is long and slender, female shape is short and broad. The pronotum has two black longitudinal stripes. Nymphs are brownish black with a pale stripe down the center; the margins of the abdomen have a light stripe.

Egg case contains 35 to 48 eggs, and hatching occurs within 24 hours after the egg case is deposited. Females produce four to eight egg cases. Development is 54 to 215 days. There are five to seven nymph stages in males and six to seven in females. Adults live about 200 days.

Habits. Survival without food or water is about eight days for males and 12 days for females; survival with water is 10 days for males and 42 days for females. Females preparing to produce an egg case leave the harborage to feed for about five days. Females carrying an egg case remain in or forage close to the harborage, but males and large nymphs forage long distances. Sticky traps containing females with egg cases indicate that an infested harborage is close by; sticky traps with males and large nymphs are probably not near an infested harborage.

ASIAN COCKROACH

Adults are about 1/2 inch long; the body is light brown or yellowish brown. This species closely resembles the German cockroach, but it is capable of flying. Nymphs are blackish brown, and the margins of abdominal segments are pale brown.

Egg case contains 38 to 44 eggs; incubation is about 20 days. Females produce six egg cases. Nymph development is about 68 days. Adult males live about 45 days and females about 103 days.

Habits. They occur outdoors in vegetation; adults fly to reflected light. Flights occur at sunset and when winds are light.

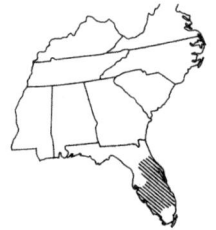

Asian cockroach distribution

FIELD COCKROACH

Adults are yellowish brown with a dark brown to black region between the eyes that extends to the mouthparts. Longitudinal stripes on the pronotum are blackish brown. Nymphs are pale yellow; large nymphs are yellowish orange.

Egg case is light brown with distinct indications of the egg compartments. Hatching is in about 20 days; females produce about eight egg cases. Development takes 45 to 56 days. Adults live 100 to 150 days.

Habits. This cockroach occurs around buildings and moves indoors during dry weather. It is active during the day, but also is found around streetlights at night.

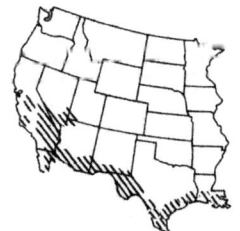

Field cockroach distribution

PENNSYLVANIA WOODS COCKROACH

Adult males are slightly more than one inch long, and females are slightly less than an inch long. Males and females are light brown, and the thorax and front wings have pale brown margins. The male's wings extend to the tip of the abdomen; female wings are small pads. Only the males fly. Nymphs are dark brown.

Egg case is yellowish brown and contains 32 to 36 eggs; about 26 eggs hatch. An egg case is produced every five to nine days and is carried for about three days; a female can produce 30 egg cases in her lifetime. Nymphs hatch in summer and complete development to adult in the spring of the following year.

Habits. This species occurs in woodpiles and accumulated forest debris in eastern U.S. Males can fly more than 100 feet; they are attracted to lights at night in May and June. These cockroaches rarely persist indoors, they do not infest houses.

Pennsylvania woods cockroach

Pennsylvania woods cockroach egg case

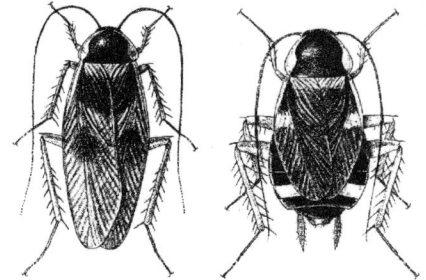

BROWNBANDED COCKROACH

Adult males are about 1/2 inch long, and females are slightly smaller. The body is brown to yellowish brown with distinct pale brown banding. Nymphs are banded light and dark brown. Adult males fly when disturbed, but females do not fly.

Egg case is brown to reddish brown; it is curved, and there are indentations showing the position of the eggs. It contains 14 to 18 eggs; hatching occurs in about 96 days. Females produce 10 to 20 egg cases. The egg case is deposited 24 hours after it is produced, and is glued to a substrate. Nymph development is about 114 days for males, and about 69 days for females. Males live about 115 days and females live about 90 days.

Habits. This species is common on furniture and in locations high on walls. Egg cases often are deposited at the same location by many females. These cockroaches have small pads on their feet, and they are not easily captured in sticky traps.

Brownbanded cockroach

Brownbanded cockroach egg case

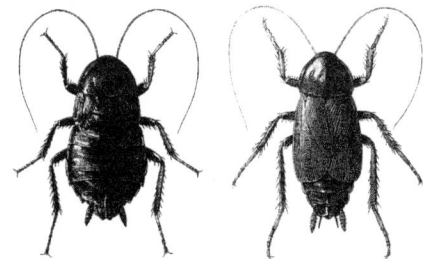

ORIENTAL COCKROACH, WATERBUG

Adult males and females are slightly more than one inch long. Body is shiny blackish brown. The male's wings cover two-thirds of the abdomen, female wings are short. Neither gender is capable of flying. Nymphs are reddish brown.

Egg case is blackish brown; it contains 16 to 18 eggs; hatching occurs in about 42 days. Egg cases do not survive when exposed to freezing. Females produce egg cases at intervals of one to two weeks; the lifetime total is six to eight. Nymph development takes 515 days for males and 542 days for females. Adults and large nymphs are active from May to early July when adults and nymphs move indoors. Adults die in July or August of the second year.

Habits. Adult females and nymphs have tarsi with only a small pad between the claws; they have difficulty climbing smooth surfaces.

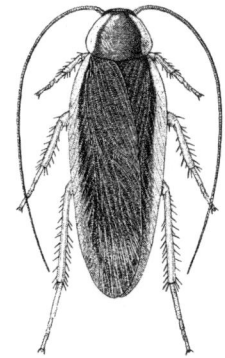

Oriental cockroach female and male

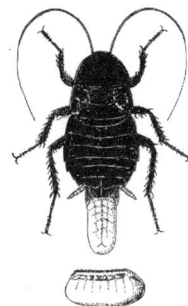

Oriental cockroach with egg case

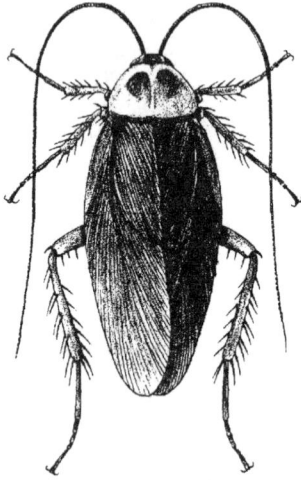

AMERICAN COCKROACH

Adult males are slightly more than two inches long and female are about one and one-half inches long. Body is shiny, reddish brown to brown; pronotum has a yellowish white margin with dark brown interior. Wings extend beyond the abdomen in males, and as long as the abdomen in females.

Egg case is dark brown to blackish brown and contains 14 to 165 eggs; hatching occurs in 57 days. Females produce 15 to 90 egg cases in a lifetime; typically 10 to 15 within 10 months. Nymph development is five to 15 months. Adult life span is 90 to 706 days for females, and 90 to 362 days for males.

American cockroach thorax

Habits. Survival without food or water is about 29 days for males and 42 days for females; survival with water is 43 days for males and 90 days for females. Adults readily fly when the temperature is above 72° F; they usually travel short distances, but sustained flight is possible. They fly to lights at night.

This cockroach occurs in urban landfills, wastewater treatment plants, and underground sewer systems of cities around the world. In homes and buildings, it is in basements and upper floors of large buildings. The female often will chew a small hole in a soft substrate to deposit her egg case, and then cover the egg case with chewed debris so that it is partially concealed. Females sometimes will eat egg cases they find in the habitat.

Seasonal abundance. American cockroach populations have a distinct seasonal abundance. Egg cases are deposited in spring; then from April through July, there are nearly equal numbers of adults and nymphs in the population. From August through November, the number of adults decreases and the number of nymphs increases. There is relatively little foraging and feeding of adults and nymphs during winter. This seasonal abundance and foraging pattern seems to be adopted regardless of the temperature conditions in the habitat.

AUSTRALIAN COCKROACH

Adult males and females are slightly less than one and one-half inches long. Body is dark brown, and the pronotum has pale margins and a dark brown interior. Large nymphs are dark brown with pale yellow spots on lateral margins of the thorax and abdomen.

Egg case is blackish brown and contains 24 eggs; hatching occurs in about 40 days. Egg cases are produced at about 10-day intervals, and the total is 20 to 30 in a lifetime. Nymph development is 6 to 12 months. Female development is about 213 days, and males about 198 days. Adults live about 12 months.

Australian cockroach thorax

Habits. The Australian cockroach is found outdoors around the perimeter of buildings and indoors in kitchens. It occupies similar habitats as the American cockroach, but does not occur in underground sewers.

American cockroach

American cockroach egg case

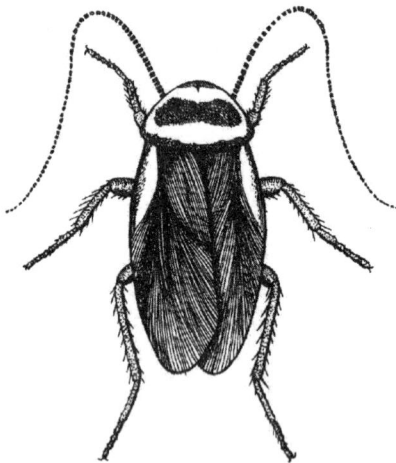

Australian cockroach

Australian cockroach distribution

TURKESTAN COCKROACH

Adult males are about one inch long and pale brown. The wings have pale yellow margins and extend beyond the abdomen. The females are dark brown with short wings. Antennae are longer than the body.

Egg case is about 3/8 inch long, is brown, and contains about 18 eggs. The nymph has a light brown thorax and dark brown abdomen.

Habits. Turkestan cockroach populations are found outdoors in the southwestern U.S., but they also invade houses and other structures. These cockroaches cannot climb. They occur in leaf litter, potted plants, and sewer systems.

Turkestan cockroach distribution

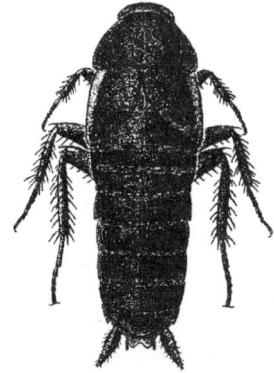

BROWN COCKROACH

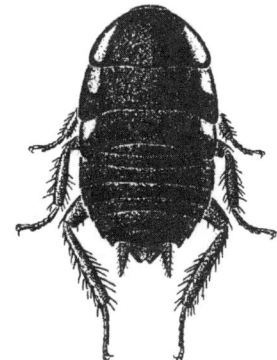

Adults are about one and one-half inches long, dark brown to reddish brown; the markings on the pronotum are pale brown. Wings cover the tip of the abdomen in both sexes.

Egg case is brown and contains about 24 eggs. Females can produce about 30 egg cases, but many are not viable. Egg cases often are partially covered with pieces of debris. Nymph development is about 263 days for females and 268 days for males. Adults live about eight months or up to 20 months, depending on environmental conditions.

Habits. This cockroach occurs primarily indoors, but also lives outdoors around trees and in sewers.

Brown cockroach distribution

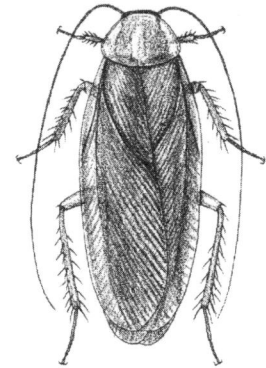

FLORIDA WOODS COCKROACH

Adults are about one and one-half inches long, and are shiny dark brown to reddish brown. Adults have small wings that are almost indistinct. Nymphs are reddish brown and have pale yellow margins on the thoracic segments; the abdomen is uniformly dark brown to blackish brown.

Egg case is brown and contains about 24 eggs; egg cases are often glued onto surfaces outdoors.

Habits. This cockroach occurs primarily in natural locations, but also in residential landscaping that includes dense vegetation. It enters houses but breeds outdoors. Adults can give off an offensive odor when disturbed.

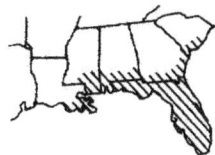

Florida woods cockroach distribution

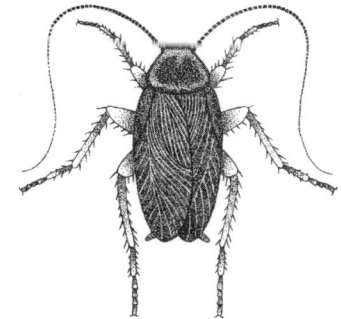

Turkestan cockroach female

Turkestan cockroach male

Brown cockroach

Florida woods cockroach

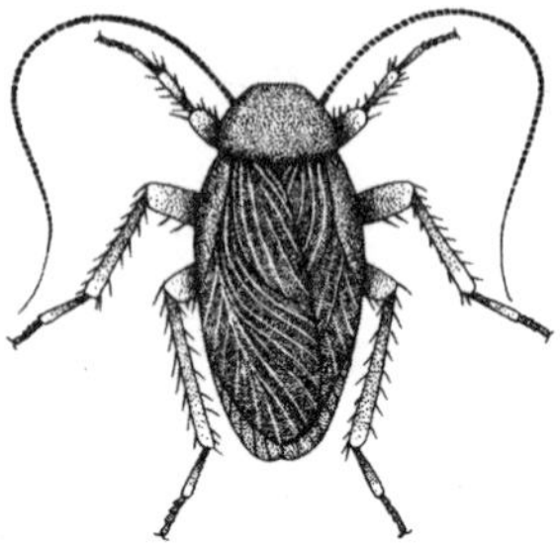

Smokybrown cockroach

Surinam cockroach nymph

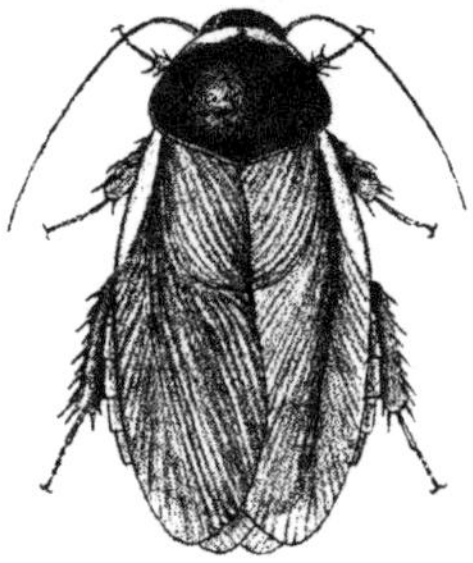

Surinam cockroach adult

SMOKYBROWN COCKROACH, PALMETTO BUG

Adults are about one and one-half inches long and uniformly dark brown to blackish brown. Wings are fully developed in both sexes. Large nymphs are uniformly reddish brown.

Egg case is brown and contains 20 to 28 eggs; hatching occurs in about 100 days. Females produce 15 to 20 egg cases in a lifetime. Nymph development is 274 to 439 days. Adults live 18 to 24 months. Adults and nymphs can withstand cold and overwinter in protected outdoor sites.

Habits. These cockroaches are strong flyers, even the females that are carrying an egg case. Outdoor habitats are moist, shaded sites, such as in tree holes and beneath the bark or bracts of palm trees, which is the basis of the name: palmetto bug.

Smokybrown cockroach distribution

SURINAM COCKROACH

Adults are about one inch long; the body is dark and shiny brown to blackish brown. Pronotum is blackish brown, anterior margin is pale brown; wings are light brown. Wings extend to the tip of the abdomen; antennae are about one-third the length of the body. Nymphs are dark brown.

Egg case contains 14 to 48 eggs, and it is carried internally until the eggs hatch in about 35 days. Development is 127 to 184 days. Adult females live about 307 days.

Habits. It occurs in greenhouses, and occasionally around potted plants in shopping malls, hotel lobbies, and similar sites. It occurs along the Gulf coast and into coastal South Carolina and North Carolina. It is established in Hawaii.

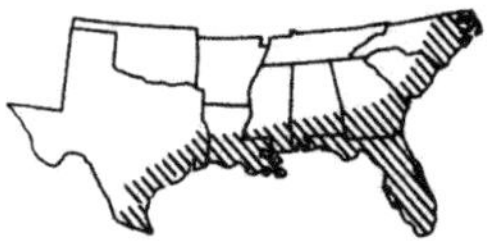

Surinam cockroach distribution

_____

_____

_____

_____

_____

_____

_____

_____

_____

_____

CHAPTER 7

CRICKETS, EARWIGS, AND SPRINGTAILS

Crickets have chewing mouthparts and are primarily plant feeders. Development is gradual; the nymph stages resemble adults except for wings, when present. Egg and nymph stages survive dry seasons or overwinter. The house cricket is the only species in this group that lives and reproduces indoors. Other crickets may be attracted to lights at night, and forage or harbor in homes, but they cannot survive long in the low relative humidity of houses. Thus, while house crickets may damage materials, other species are only a nuisance. Large numbers of field crickets can be a nuisance in late summer when they move to the perimeter of buildings during cool nights, then move inside the houses and building around doors and windows. These crickets also are capable of flying.

Earwigs are slightly flattened insects with a pair of movable forceps at the end of the abdomen. The forceps are large in males and small in females. Earwigs have gradual metamorphosis and chewing mouthparts. They are nocturnal, and they feed on plant and animal material. There are four to six nymph stages, and adults appear in late summer; they overwinter as adults. Earwigs are attracted to lights at night and enter buildings around doors and windows. They often gather in narrow harborages. The name "earwig" is based on the superstition that they crawl into people's ear at night. The straw used in early bedding provided harborage for earwigs, and occasionally one of these insects would be found in the ear of a sleeper.

Springtails are wingless, soft-bodied insects. They have chewing mouthparts, and the metamorphosis is simple; the immature stages resemble the adults. The common name is derived from a tail structure that they use to propel themselves through the air. Their jumping ability helps them escape from predators. Springtails inhabit moist locations, and most feed on decaying plant material, fungi, pollen, and algae. Those that occur indoors usually are associated with moist or wet conditions, but some species can persist in dry environments. Favorable indoor conditions include high humidity, mold, or other wet or moist organic matter.

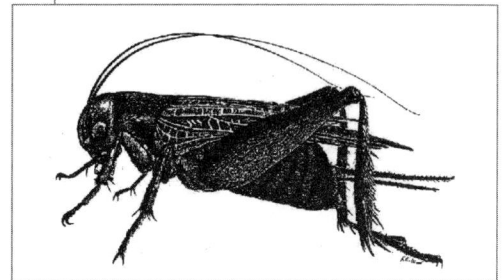

Field cricket male

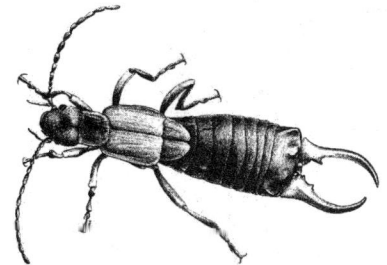

Earwig

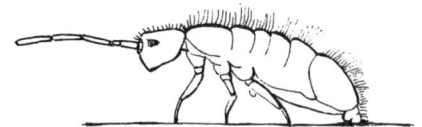

Springtail sitting

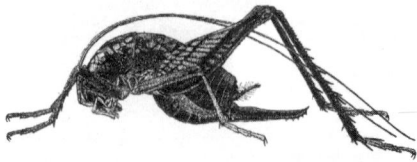
Camel cricket

Jerusalem cricket

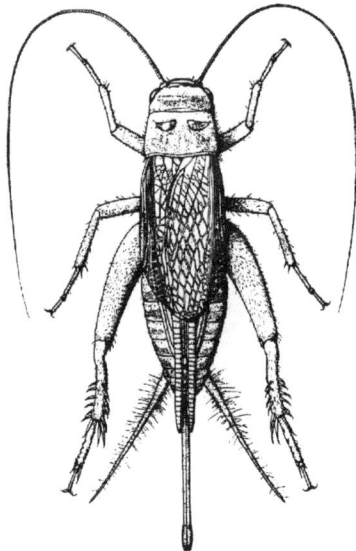
House cricket

CAVE CRICKETS, CAMEL CRICKETS

Adults are about one inch long; the body is light brown to dark brown and may have a mottled pattern on the thorax and abdomen. Antennae are longer than the body.

Development is about 64 days for males and females; there are about six nymph stages. Adults live about 90 days. They overwinter as immatures and adults, with one generation per year.

Food is primarily plant material, including fungi, decaying leaves, and roots in crawlspaces.

Habits. Camel crickets do not chirp. The common names of these two crickets refer to their habits: the cave cricket is found in dark habitats, and the camel cricket has a high arched thorax.

JERUSALEM CRICKETS

Adults are about three inches long; the body is light brown to dark brown. They are wingless, and the legs have sharp spines.

Development is through nine to 11 molts and is completed in about 18 months. Adults appear in midsummer. Females dig a small hole in the soil to lay eggs, which hatch in the fall or spring. There is one generation per year.

Food is primarily plant material, including roots and tubers; they also may feed on dead animals.

Habits. These large crickets often are found around the perimeter of buildings, and sometimes occur indoors and in swimming pools. The human-like head of the adult has created superstitions around these crickets. In southwestern United States and Mexico they are called *nina de la tierra* or child of the earth.

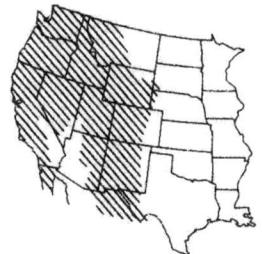
Jerusalem cricket distribution

HOUSE CRICKET

Adult body is yellowish to mottled or pale brown with three dark cross-bands on the head. Eggs are deposited singly or in small batches in moist cracks and crevices; females can lay 40 to 179 eggs. Hatching occurs in one to 12 weeks.

Development is about 56 days for males and 53 days for females; there are nine to 11 nymph stages. Adults live about 90 days. They overwinter in the egg stage, with one generation per year.

Food is primarily outdoor plant material; inside homes they feed on foods.

Habits. Adults are attracted to lights at night; they can climb rough-surfaced buildings. They can occur in large numbers in refuse dumps and urban landfills, which are usually kept warm throughout the year by the fermentation of the wet, organic material. From landfills, crickets can then move to surrounding buildings.

FIELD CRICKETS

Adult body color ranges from black to yellowish brown; front wings can have orange markings. Eggs are deposited singly in damp soil; females can lay 150 to 400 eggs.

Development takes 78 to 93 days; there are eight to nine nymph stages. Adults live for about two months; in the fall, they are usually killed by frost.

Food is plant material, but indoors they damage fabrics, such as cotton, wool, silk, and fur. There are one or two generations per year, and they overwinter in the egg stage.

Habits. These crickets are most abundant in the fall when adults gather at structures, attracted to lights or the sunlight heat retained by the structures during the day.

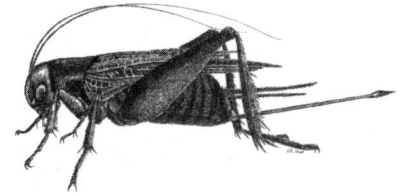

Field cricket female

COMMON EARWIG

Adult body is reddish brown; the legs are pale yellow. Males may be of different sizes, and their forceps can reach the length of 0.38 inch. Eggs may be deposited in cavities in the soil and laid in fall and/or spring to produce two generations in a year. Eggs that are laid in the fall hatch in about 73 days; eggs laid in the spring hatch in about 20 days.

Development of the four instars is about 68 days. Adults appear in late summer. Both males and females produce an aggregation pheromone, which may explain why they often are found in large groups in harborages.

Food includes a variety of green plants and insects, such as aphids, mites, insect eggs, and caterpillars.

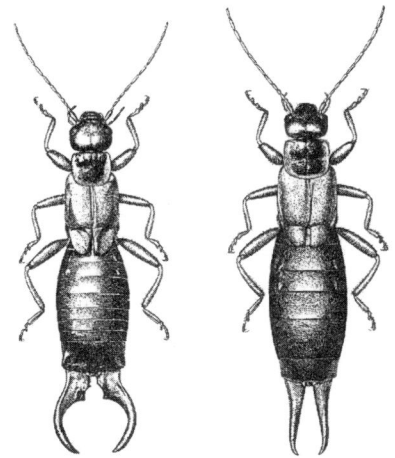

Earwigs

HOUSEHOLD SPRINGTAILS

Adult body is slender and brownish black to gray; antennae are long. High numbers and extensive infestations have been reported in domestic habitats, such as in kitchens, bathrooms, and clothes closets. Large populations also may occur around the outside of buildings. They may be spread from one location to another in household materials.

Eggs are deposited singly or in small batches directly on a moist substrate. There are six to eight molts before nymphs achieve maximum size; development is completed in about 48 days. Full-grown springtails live for about 15 days. There are multiple generations per year.

Habits. There is a tendency for gregariousness and massing of large numbers of adults and nymphs for short periods. This behavior is usually associated with abundant food, favorable environmental conditions, or migration.

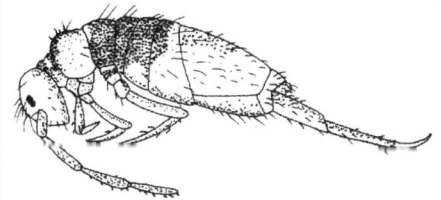

Springtail

_____

_____

_____

_____

_____

_____

_____

_____

_____

_____

_____

_____

_____

_____

_____

CHAPTER 8

FLEAS, LICE, SILVERFISH, AND PSOCIDS

Fleas are laterally compressed and wingless insects. Their mouthparts are piercing-sucking which enables them to suck blood from a host animal. Adult fleas are parasites of warm-blooded vertebrates. The majority of species occur on mammals, but about 100 species are found on birds. Fleas have complete metamorphosis, with distinct egg, larva, pupa, and adult stages.

Adult fleas remain on the body of the host and suck blood about once per hour, so they seldom leave the host animal. If they do, they quickly jump back to the host. Eggs are laid on the host, but fall to the ground. Larval stages feed and develop on the ground or bed of the host animal (or indoor carpeting); they have chewing mouthparts and resemble fly maggots. When development is complete, the larvae often will move to a location different from the larval feeding site and create a silk cocoon, which is usually covered with pieces of debris from the substrate. The pupa forms inside the cocoon, and the flea emerges as an adult by breaking open the cocoon and crawling to the surface.

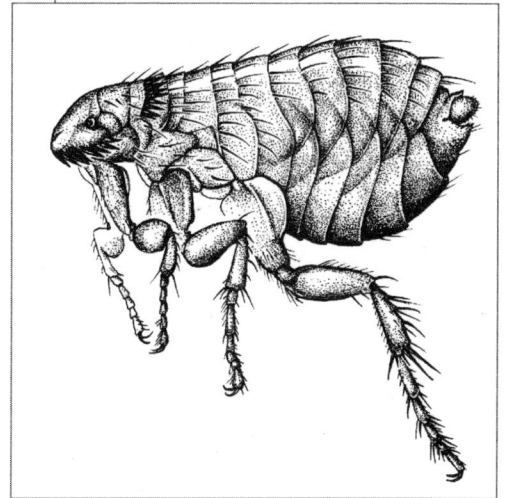

Cat flea adult

Lice are parasites of mammals or birds. Sucking lice live on the skin of mammals, and all stages suck the blood of their hosts. These insects are host specific and do not infest animals other than their primary host: the dog louse remains a pest of dogs and the human louse remains a pest of humans. The louse stays on its host throughout its life cycle; and lice almost always are transmitted by contact. Humans are hosts of head lice and pubic, or crab, lice.

The shape of the louse's claws influences its ability to infest different races of people. In the U.S., head lice infestations are 35 times higher among Caucasians than among Blacks. Contributing to this difference is the ability of the claws to grasp the hair type. Caucasian hair is round whereas Afro-Caribbean hair is flattened oval. The claws of head lice and body lice may have difficulty grasping the hair of non-Caucasians.

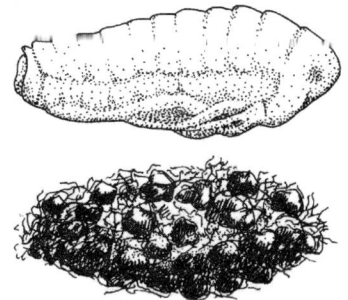

Flea pupa & case

Silverfish have flat bodies that give them access to cracks and crevices in kitchens and bathrooms, and wall voids throughout houses and commercial buildings. They are active at night and move quickly in and out of harborages, which makes infestation sites difficult to find and treat. Sticky traps can be useful in locating infested harborages, so liquid and dust applications can be more effective. Silverfish often are found in sinks and bathtubs, because they cannot climb smooth, vertical surfaces.

Psocids are common household pests; they usually occur in kitchens and bathrooms, but can be found in other humid locations. In commercial buildings, they can occur in rooms where books and files are stored. There are several psocid species that infest stored grain products, especially if the product is exposed to moisture. Inspection for these small insects is difficult without the aid of a hand lens.

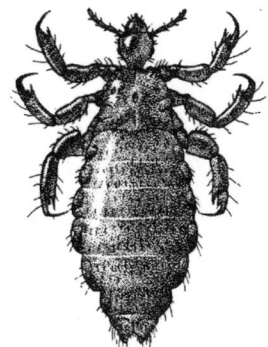

Louse

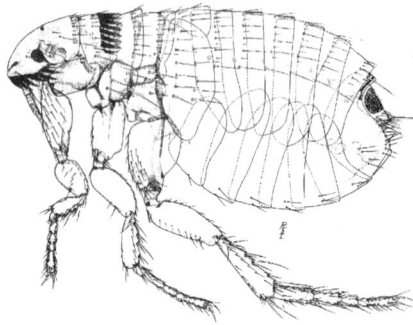

Cat flea adult

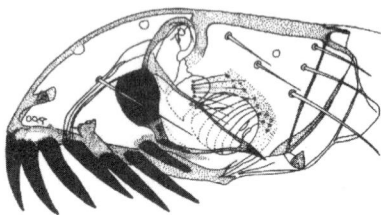

Cat flea head

Cat

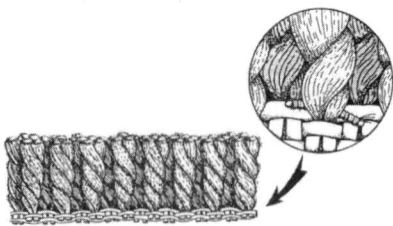

Cat flea larva in carpet

CAT FLEA

Adults are brown to yellowish brown; males are slightly smaller than females. Full-grown larvae are yellowish white, but may be reddish brown after feeding on dried blood.

Under magnification, the adult cat flea can be distinguished from other fleas found on domestic animals, including the dog flea, by its long head and seven large setae at the front. The first three setae on the head are prominent on the cat flea, but are relatively small on the dog flea.

Eggs are smooth and shiny; about 70% fall from the animal within eight hours, usually when the host shakes or scratches. Eggs are found at sites where the infested animal sleeps or rests. Hatching occurs in about 48 hours. Females lay 40 to 50 eggs per day; the total number produced is 300 to 800 eggs.

Larval food is organic matter and dried blood feces of the adult flea. Larvae complete development and pupate within 34 days of egg hatch. Emerged adults may live for about 113 days on a suitable host, but only about 20 days if a host is not available. Mechanical pressure, vibration, and heat will stimulate the emergence of adult fleas from the cocoon. Walking onto carpet infested with fleas usually provides the vibration and pressure to trigger hatching and cause the emerging adult fleas to attach to and bite ankles and lower leg regions.

Habits. The cat flea is found primarily on domestic dogs and cats worldwide, but also on feral mammals in urban areas, including opossums and skunks. The cat flea is much more common than the dog flea, and it occurs on both animals.

On animals, adult fleas are most commonly found at the base of the tail and on the head, as blood flow is close to the surface of the skin in these areas. When indoors, recently emerged fleas will attempt to feed on humans. Human skin is difficult for fleas to penetrate, but they may try several times. Flea bites on people often occur in a line of three or more bites.

Distribution. The location of cat fleas indoors is linked to the behavior of the infested pets. Sites where the animals sleep or rest have accumulations of dried blood feces from adult-flea feeding. These sites have large numbers of cat flea eggs and larvae, corresponding to where the dried blood falls to the floor.

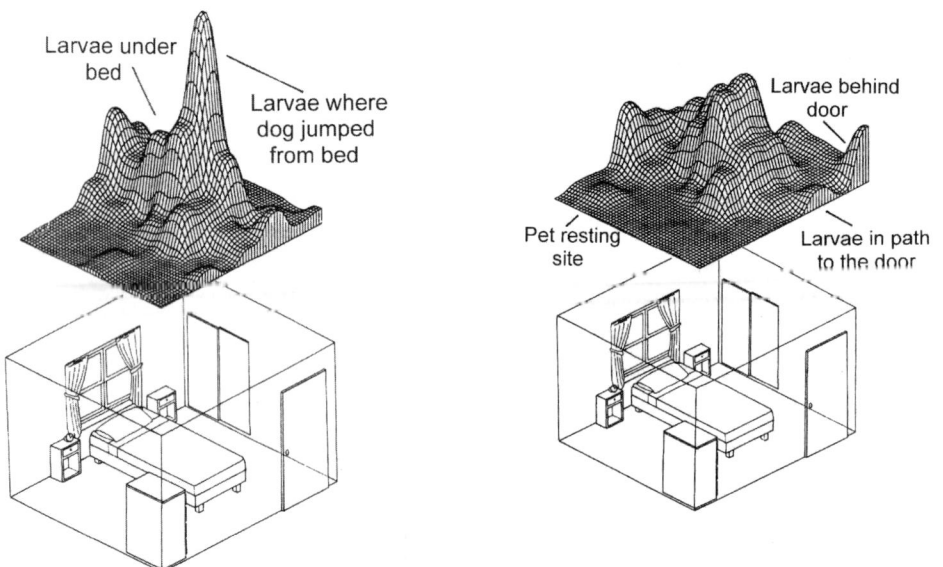

Larvae under bed

Larvae where dog jumped from bed

Larvae behind door

Pet resting site

Larvae in path to the door

Flea distribution in room

HEAD LOUSE

Adult body is gray to translucent, but usually resembles the hair color of the host. The head is short and constricted at the base; it has a short neck. Legs have well-formed claws. The male's abdomen is pointed at the tip; in females, the abdomen ends in two triangular projections. Nymphs resemble adults. Adults and nymphs live and feed on the body of the host; they are usually found on the neck and head, particularly behind the ears and on the back of the neck. Eggs are glued to hairs on the head and neck.

Eggs are yellowish white. Females attach eggs singly, close to the base of a host hair. Hatching occurs in seven to 10 days. The female lays about seven eggs in 24 hours, laying a total of about 55 eggs in her lifetime. Human scalp hair grows about 0.4 mm per day, and as it grows, the egg or nit is moved progressively farther from the scalp. Immature development is completed in eight to nine days. Males and females live about 10 days; adults and nymphs can survive about 55 hours away from the host.

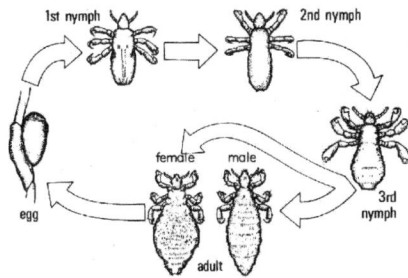

Spread is usually through the sharing of clothing or hair tools (combs, brushes, etc.) that have stray hairs with eggs or lice attached, or through close and prolonged physical contact. It is common, around the world, for school children to get lice, and this often is wrongly associated with neglect or unclean conditions at home.

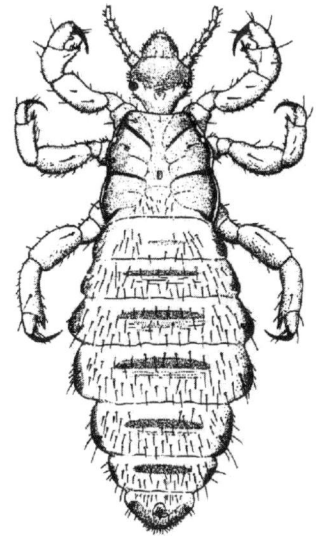

Louse life cycle

Head louse

SILVERFISH

Silverfish are slender, flattened, and tapered, with three tail-like appendages at the end of the abdomen. Their bodies are usually covered with shiny, fish-like scales—which is the origin of their common name. Mouthparts are the chewing type. Development progresses through many stages. They are long-lived insects.

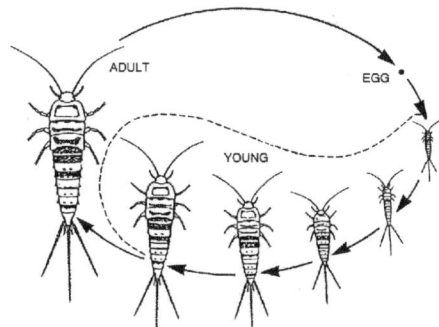

These insects are primarily nocturnal and are scavengers. Indoors, they feed on starchy material, such as old book bindings, starched clothing, and starch-based wallpaper glue. They often occur in large numbers in attics with wood-shingle roofs. Firebrats will attack knitted or plain-weave fabric. The firebrat can digest starch, fat, and protein. The four-lined silverfish can digest cellulose material.

Silverfish development

Silverfish

COMMON SILVERFISH

Adults are about one inch long, not including the long terminal appendages. They are silver-gray, with a metallic sheen.

Egg production totals about 100. Development from nymph to adult takes 90 to 120 days. Adults live about three years.

Habits. Food is usually carbohydrates and protein material. In food storage areas, they feed on flour, meal, and other similar products; cotton and silk fabrics are also attacked.

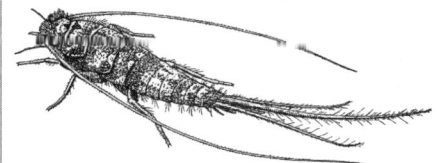

Silverfish habitat

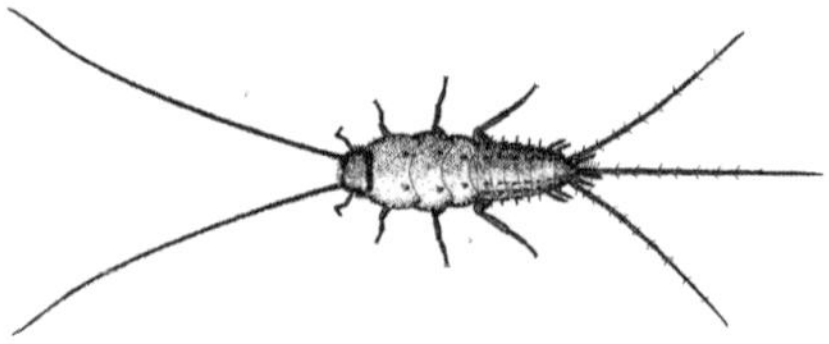

Silverfish

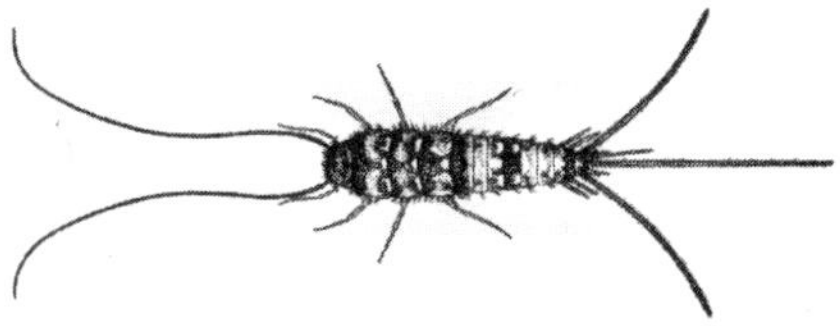

Firebrat

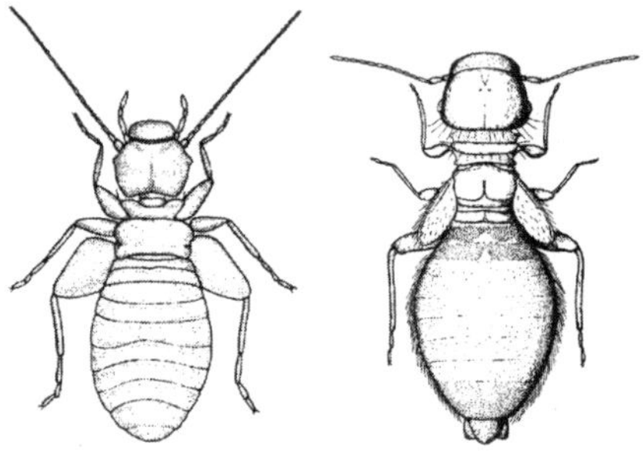

Psocid

FOUR-LINED SILVERFISH

Adults are brownish gray, with four dark lines extending the length of the body. This species can be found throughout infested structures including basements, wall voids, and attics. Infestations usually are large in houses with wood-shingle roofs.

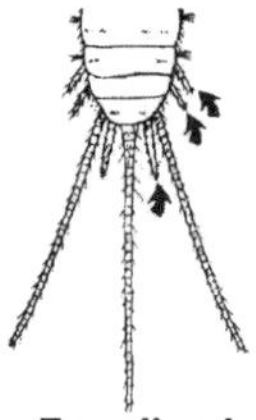

Four-lined silverfish close up

FIREBRAT

Adults are silvery gray, with a somewhat mottled gray appearance.

Egg production totals about 50 eggs. Development is through a long series of molts. There are about 13 days between successive molts, and individuals have 45 to 60 molts in a lifetime. Adults live about two years.

Habits. Firebrats prefer indoor locations with temperatures above 89°F; optimum development occurs between 98°F and 102°F. They are pests in commercial locations that maintain high temperatures, such as food processing plants, and in equipment rooms.

CEREAL AND HOUSEHOLD PSOCIDS

Adult have a brown body with striped abdominal segments.

Eggs are laid in batches of two to three a day, with females producing about 200 eggs in total. Hatching occurs in about 11 days, but eggs may overwinter. Development is through four nymph stages over a 24- to 65-day period. This species does not develop in locations with less than 55% relative humidity.

Habits. These psocids are widely distributed in households, and they probably are dispersed with the movement of food materials. They are common pests of food storage facilities and retail food stores; in both locations, infestations may occur on pallets, in packaging, and in the product. In retail stores, this species is common in flour, cereal products, and sugar.

_____

_____

_____

_____

_____

_____

_____

_____

_____

CHAPTER 9

FLIES, MOSQUITOES

Flies have well-developed front wings, but the hind wings are simply small knobbed structures. Adult flies are active during the day, sometimes at dawn or dusk. The larva is the primary feeding stage, while the adult stage is dedicated to dispersal and laying eggs. Larvae are usually buried in the food substance, with only their posterior end exposed for breathing. Some adults, such as moth flies, remain close to the larval development site; others, such as house flies, move far from that site.

Food eaten by fly larvae includes plant and animal material—some fresh, some partially decayed, and some rotting. Food location and selection is conducted by the female; she selects material that will be suitable for the larval stage. Females are usually attracted to odors, because the gases indicate a suitable larval food source. Fruit fly females will return to the site of their larval development to lay eggs for the next generation. Blow flies and phorid flies can locate and deposit eggs on the carcass of a dead rat or mouse deep in a wall void.

The larva or maggot is the primary feeding stage, and egg-to-adult development usually takes about 10 days. A fly larva develops into an adult in a protective case called a puparium. The adult emerges from the puparium in about 10 days by pushing the end of the puparium open. Adults usually do not feed or they take only liquids, and they do not live long. For example, an adult house fly or fruit fly lives about 30 days; an adult midge may live only a few hours or days. The female mosquito must take a blood meal to live and develop her eggs.

Fly control in and around restaurants and kitchens begins with inspection and identifying of larval breeding sites. The breeding sites for most fly pests are outside, but some may be inside. House flies breed primarily outdoors in decaying garbage; moth flies usually breed in indoor decaying organic material. Mosquitoes breed in standing water that may occur in discarded containers or small pools. Female mosquitoes are capable of flying a long distance to find a blood meal. The Asian tiger mosquito frequently breeds in small amounts of water around buildings. This species is active in late spring and throughout the summer; sometimes there is a second peak of activity in the fall.

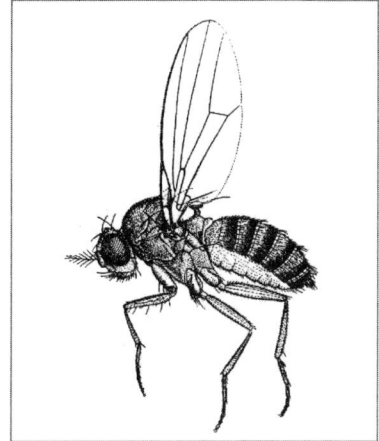

Fruit fly adult

Fly maggot

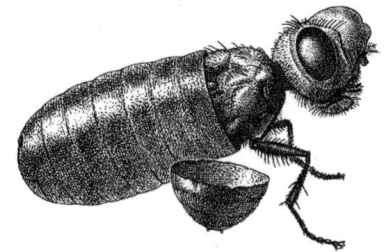

Fly emerging from puparia

Yellow fever mosquito

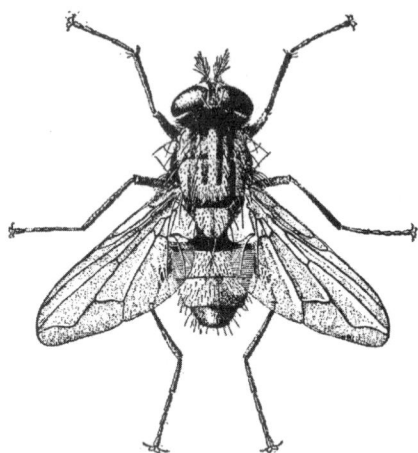

HOUSE FLY

Adults have four light stripes lengthwise on the thorax. Full-grown larvae are yellowish white.

Development. Eggs are deposited in batches of 75 to 150; females may deposit as many as 21 batches of eggs for 31 days. Eggs hatch in eight to 12 hours. Larval development is completed in about five days, and full-grown larvae move to a dry substrate before pupating. Adult flies live about 30 days during warm months, but this may extend to 60 days.

Habits. The usual distance traveled by adult house flies is about 1,300 feet. Adults are most active between about 2:00 and 4:00 p.m, which is the hottest and driest time of day. Adults are attracted to artificial light during the day or night.

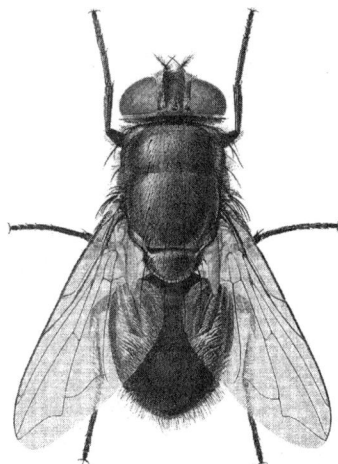

BLOW FLIES

Blow flies are large blue- and green-colored flies found in residential and commercial kitchens and garbage and dumpster stalls. This fly is easily recognized by its bright shiny color, which is usually uniform for the entire body; it has no stripes on its back.

All blow fly species have the same basic life cycle: egg, three larval instars, a brown puparium containing the pupa, and eventually adult. They complete their life cycles in 10 to 20 days, depending on temperature. These flies develop in decaying organic matter and substrates with high protein content. Decomposing animal carcasses are a common breeding site.

BLUE BLOW FLY

Adults have a bluish-black thorax, and a metallic-blue abdomen.

Development. Eggs are laid in batches of up to 180 directly on the larval substrate; females can produce 500 to 700 eggs. Hatching is in about 20 hours. Larval development is completed in about three days during the summer. Adults live about 30 days. Females lay eggs on fresh, decaying, or cooked meat, and on human excrement.

Habits. Adults appear in early spring, but there may be a peak of adults in fall because the adults that emerged in late summer will live for about 30 days. These are slow-flying and loud-buzzing flies, and they often enter houses.

GREEN BLOW FLY

Adults are metallic green. Full-grown larvae are about 1/2 inch long and may be colored slightly purple.

Development. Eggs are deposited in large numbers in one location; females can lay a total of about 250 eggs. Hatching occurs in about 24 hours. Larval development is completed in about five days.

Habits. This fly breeds in dead animals, decaying garbage, and manure. Green blow flies are common in and around houses, and adults can enter openings as small as 1/8 inch. Larvae will develop in dead birds or rodents in attics and wall voids.

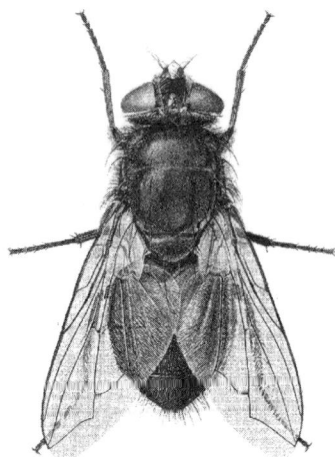

House fly

Blow fly

Blow fly

CLUSTER FLY, ATTIC FLY

Adults have a broad thorax covered with golden-yellowish setae; the wings overlap when at rest.

Development. Eggs are laid singly in the soil, and hatching occurs in about three days. Larvae are predaceous on earthworms, with first-stage larvae seeking out earthworms and entering their body. Larval development is completed in 27 to 39 days, and the puparium is formed in the soil. There may be four generations per year.

Habits. In late summer and fall, large numbers of adults gather on the sun-warmed sides of buildings, and then move through cracks and crevices to enter attic spaces and wall voids. They remain there throughout the winter. On warm and sunny days of early spring, adults leave; they mate in the spring.

FRUIT FLIES

Fruit flies gather around ripe and decaying fruit and decaying vegetation indoors and outdoors. The small size of the adult gives it access to food sources that are unavailable to some other flies, but its weak flight limits its activity to protected locations. There are about 10 species of fruit fly associated with man-made habitats. The most common are *Drosophila melanogaster,* which is a fruit-feeder in the larval stage; *D. funebris* breeds in organic waste, including feces.

Adults are attracted to a variety of organic compounds that are found in fermenting material. Under ideal conditions adults live about 40 days; however, some species may live only seven days. Adult females begin laying eggs two days after emergence. Egg production increases to a maximum on the fifth day, and then is maintained for three to 10 days. Females lay eggs directly on the larval feeding substrate, which may be wet or liquid. Fruit-fly eggs are equipped with floats that keep them from sinking into the substrate. The posterior breathing openings at the end of the larva are located on an extension. This keeps the breathing pores in contact with the air so that the larvae can remain immersed in the substrate.

RED-EYE FRUIT FLY

Adults are yellowish brown, with bright red eyes. Larvae are 1/4 inch and pale yellow.

Development. Eggs are laid directly on a substrate; the female lays about 600 eggs in her lifetime. Hatching occurs in 12 to 24 hours. Larval development from egg to adult is eight to 10 days. Adults emerge in about four days, and live about 30 days in humid conditions.

Habits. Larvae feed in decaying fruit and vegetables. The larva's posterior breathing tube is short; it does not feed deep below the surface of the food substrate.

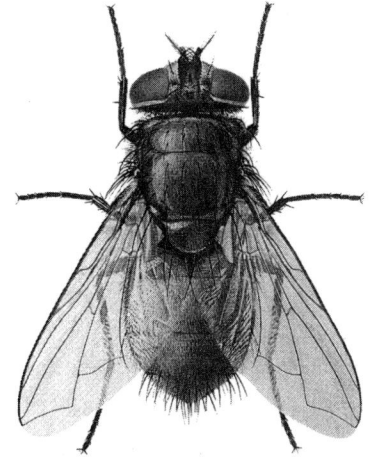
Cluster fly

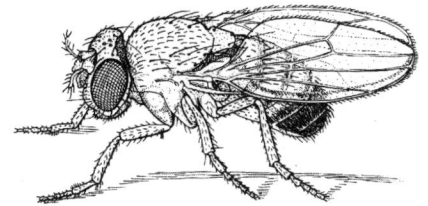
Red-eye fruit fly

Fruit fly larva

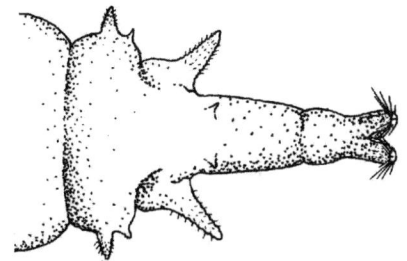
Fruit fly larva breathing tubes

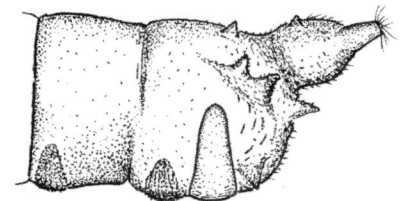
Red-eye fruit fly larva tail

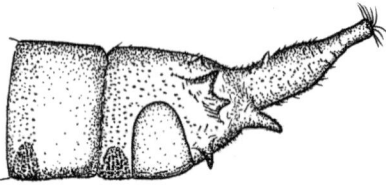

Dark-eye fruit fly larva tail

DARK-EYE FRUIT FLY

Adult abdomen is dark brown. Larvae are about 1/4 inch long, yellowish white.

Development. Eggs are laid directly on the substrate. Hatching occurs in 12 to 24 hours. Larval development takes 13 to 15 days. Indoor populations may persist year round.

Habits. Larvae feed in decaying organic matter. The posterior breathing tube on larva is long. Larvae feed deep below the surface of the food substrate, and the long breathing tube allows them to do this.

PHORID FLY

Adult body is yellowish brown; the abdominal segments are yellowish brown. Full-grown larvae are pale yellow. The puparia are brown and have two distinct "horns" on the anterior surface.

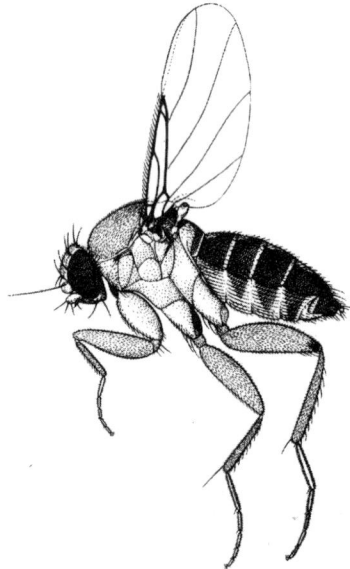

Phorid puparium

Development. Eggs usually are laid at the edge of the substrate, and females lay eggs for about 30 days. The total number of eggs laid is about 600, but can be as many as 1,000. Hatching occurs in about 24 hours. Larval development is complete in about three days. Total development time is about 13 days.

Habits. Adults are found close to the breeding site and often at windows and lights near the site. This species feeds in decaying plant and animal matter, including sewage and household organic waste. Phorids are often found in hospital and health care facilities, where they infest organic substrates, such as human waste.

Phorid fly

FUNGUS GNATS

Adults have a black body; the wings are nearly black. Larvae are white and slightly transparent, the head is black.

Fungus gnat larva

Development. Eggs are deposited in crevices in batches of up to 30 eggs; females lay a total of about 175 eggs. Hatching occurs in about seven days. Laval development is completed in two to three weeks. Adults live about 10 days and feed on moisture in the soil.

Habits. Fungus gnats occur indoors in the moist or wet soil used for potted houseplants. Mating is on the surface of moist, organic substrates; there are no mating swarms.

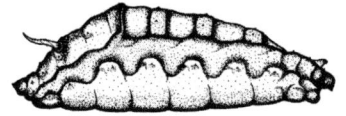

Fungus gnat

MOTH FLY, DRAIN FLY

Called both moth fly and drain fly, these adults are uniformly gray and covered with fine setae or hairs. Full-grown larvae are yellowish white to light brown. There are dark areas on the dorsum of all the larval segments. The larva head is dark brown and the last segment of the body is elongated and dark brown to black.

Development. Eggs are laid in batches of 20 to 100 directly on decaying substrates. Hatching occurs in about 48 hours. Larval development is complete in nine to 15 days. Adults are weak flyers; indoors they rest on walls close to the breeding site. They are not attracted to lights at night and may not come to UV light traps.

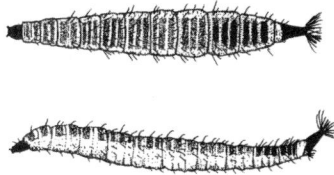

Habits. Females search for egg-laying sites. The most common sites are clogged drains and the organic material around fixtures in bathrooms and kitchens. Treating the clogged drain with various chemicals rarely controls these flies. The best control strategy is physical removal of the clogging material, and then maintaining the drain with biocleaners.

Drain fly larva

Drain fly adult

RED-TAILED FLESH FLY

Adult body is blackish gray with three dark stripes on the thorax and a black-and-gray checkerboard pattern on the abdomen.

Development. Eggs develop within the female's body, and she is able to deposit first-stage larvae directly on substrates. Larval development is completed in about six days, and the pupal period lasts eight to 10 days. Adults live about 30 days; there are several generations per year.

Habits. Large numbers of adults may occur indoors if larvae have been feeding on the body of a dead animal in a wall void or chimney.

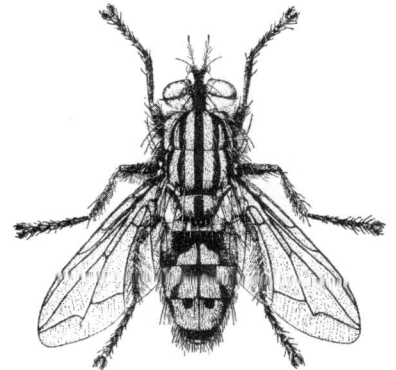

Flesh fly adult

CRANE FLIES

Adults are brown to grayish brown. Their wings are mottled brown and usually slightly transparent. The adults do not feed and live for only a few days.

Development. Larval development is in leaf litter and wet organic material; some species breed in the thatch layer in turfgrass. Adults emerge in spring and begin laying eggs soon after they fly. There are two or three generations per year; adults are most often found in spring or early summer.

Habits. These flies are common in urban and rural areas; they are sometimes considered to be large mosquitoes, but they are harmless. These flies often come to lights at night and will occur indoors around lights.

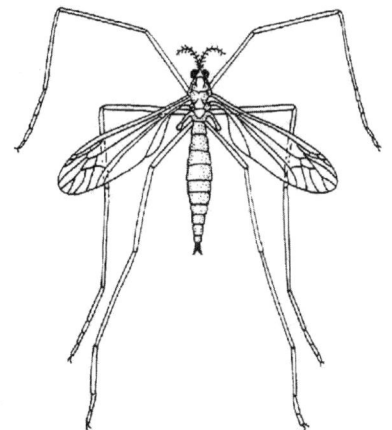

Crane fly

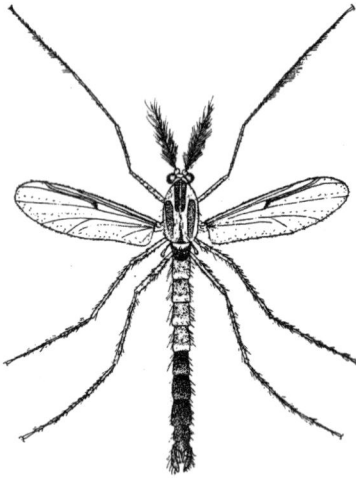

Midge adult

Midge larva

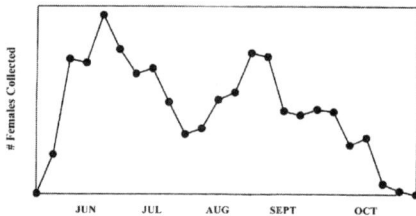

Asian tiger mosquito population trends

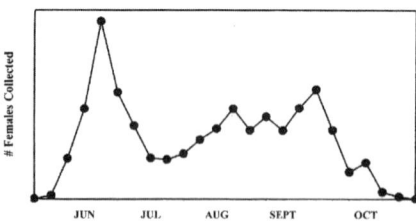

Flood water mosquito population trends

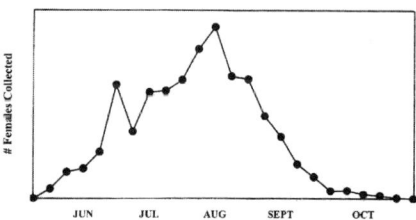

House mosquito population trends

MIDGES

These flies are about 1/2 inch long and have gray to black bodies. The wings are clear and the antennae are large and bushy.

Development. Larval development is in ponds, lakes, and slow-moving rivers, where the larvae feed on plant material. Larvae complete development in about three weeks, and the adults emerge to fly for only about four days. The adults do not feed, but die soon after mating and laying eggs.

Habits. Swarms of these delicate flies are common in spring and fall. They are often attracted to lights at night, and they sometimes fly indoors. They may be mistaken for mosquitoes, but adult midges have no mouthparts and cannot bite.

MOSQUITOES

Adult males remain close to the breeding site and do not bite; females may fly long distances to find a blood meal. They seek wet or aquatic sites to lay eggs. Larval stages are aquatic; the pupal stage also remains in the water.

Development depends on temperature, and ranges from seven days to seven months. For most species there are two or three generations per year. Some overwinter as unfed females, some in the egg stage.

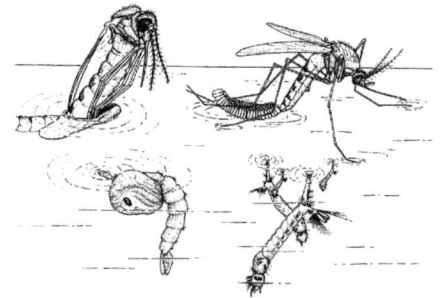

Mosquito life history

ASIAN TIGER MOSQUITO

This is a day-biting mosquito, with an early morning and late afternoon peak. Females can travel a mile or more from the breeding site to find a blood meal. This species breeds in containers and small collections of water around buildings.

FLOOD-WATER MOSQUITO

This is a day-biting pest. Females can move more than 20 miles from the breeding site. They breed in woodland pools. In urban areas, this mosquito is abundant in late summer; it moves indoors to bite.

HOUSE MOSQUITO

This is the most common household mosquito species. Larvae live in a variety of urban habitats, including street drains, gutters, domestic containers, and drains. This species actively enters houses and buildings to bite humans.

CHAPTER 10

MOTHS

The caterpillars of several moth species are pests of wool and silk, while other species infest dry foods, such as flour and cereal products. It is the caterpillar stages that feed on and damage products; the adult moth searches for new sites to lay eggs and extend the infestation. The Indian meal moth is one of the most common and widespread household pests because the caterpillar stage feeds on nearly every food in kitchen cabinets, as well as on dry pet and bird food.

Eggs of moth pests usually are deposited directly on the food source of the caterpillar stages. Hatching occurs in several days, and caterpillar development is completed in one to three weeks. The number of caterpillar stages ranges from five to eight, depending on the species and available food.

The full-grown caterpillar moves away from its feeding site to form a cocoon and then a pupa. Caterpillars have silk glands that open at their mouths; they use silk to make feeding shelters and to wrap and protect the pupal stage. Pupae of most species are encased in a silken cocoon; in some species the caterpillar builds a pupal case from the material it is infesting.

Pest status for members of this group is based primarily on the damage each does to stored food, fabric, and other materials. Several species have adapted to indoor habitats and to the food and fabric stored there. The caterpillars of stored food pests can penetrate the seams and small openings in modern packaging material. The species that attack wool and silk fabric are easily transported to other locations in infested material. Most of the flour and grain pests have been distributed around the world with commercial shipments of food and materials.

Moth pests of stored food can be detected with sticky traps that use a pheromone as an attractant. Most traps target a single species, but some use pheromones formulated to attract several closely related species. Pheromone traps can provide information on the species involved, seasonal prevalence, locations infested, and the presence of the adult stage in the population cycle. Pheromone traps release a female sex pheromone or male aggregation pheromone that attracts only males in the vicinity. Males detect low concentrations of this chemical and fly toward the source.

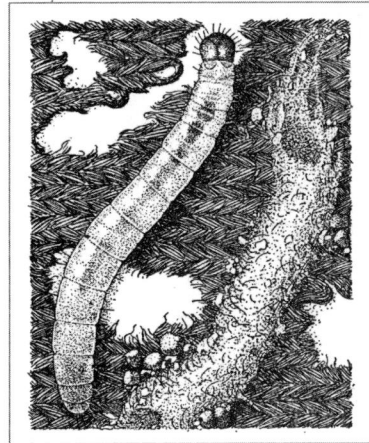

Clothes moth caterpillar

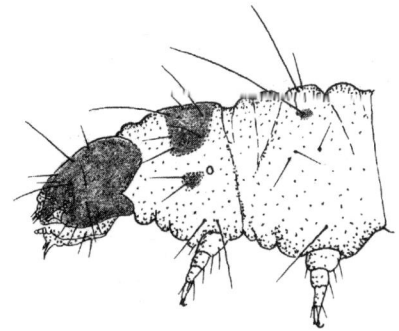

Indian meal moth caterpillar head

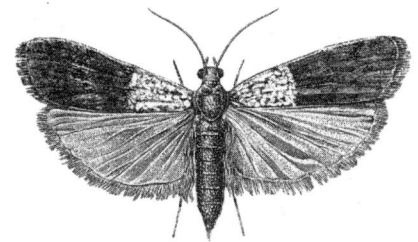

Indian meal moth

Corn damaged by moths

Indian meal moth

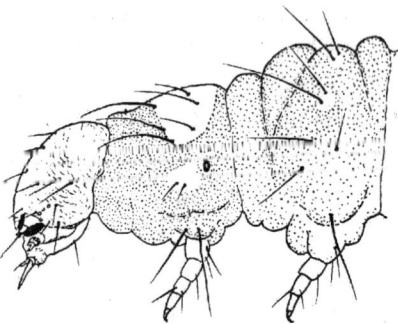

Indian meal moth caterpillar

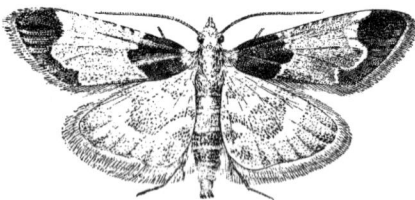

Meal moth

Meal moth caterpillar

ANGOUMOIS GRAIN MOTH

Adult wingspan is about 1/2 inch; the body is grayish to yellowish brown. The full-grown caterpillar is about 1/4 inch long; its body is white with a yellowish-brown head.

Development. Eggs are laid directly on the surface of food; a female can produce 80 to 200 eggs in her lifetime. Hatching is in about 10 days. Caterpillar development is completed in about 21 days, and total development time is about 40 days. There are four or five generations per year, but in heated buildings there may be 10 to 12 generations.

Habits. Caterpillars attack stored whole grain or caked grain in containers. In households, ears of ornamental corn (Indian corn) may become infested. The adults typically remain near infested material and will take flight in large numbers when the material is disturbed.

INDIAN MEAL MOTH

Adult wingspan is about one inch; the wings are pale gray, and the outer portion of the forewing is reddish brown. Full-grown caterpillars are about 1/2 inch long and yellowish white, but may be greenish or pinkish white.

Development. Eggs are deposited in the food; a female lays 150 to 400 eggs in her lifetime. Hatching is in four to five days. Caterpillar development is completed in about 60 days. Full-grown caterpillars move away from the infested site to pupate; they wander for many hours and long distances before stopping to make their cocoons.

Habits. Adults are weak flyers and remain at rest for long periods on walls. Caterpillars are often found crawling on walls and ceilings near or some distance from the infested material. They are often misidentified as maggots. The adults may remain close to the infested site; they typically fly short distances and then rest. They are not attracted to lights at night.

MEAL MOTH

Adult wingspan is about one inch. The front wings are light brown in the middle and dark brown at the base and the tip. Caterpillars are about one inch long; the head is black, and the posterior is pale orange.

Development. Eggs are scattered on the food surface; females lay 200 to 400 eggs. Caterpillar development is completed in about two months. Caterpillars feed from tubes of silk, which contain particles of food.

Habits. The caterpillars feed on flour, meal, damaged grain, seeds, sesame, peanuts, and vegetable refuse. This moth is common in flour processing and storage facilities. It often collects in UV light traps.

MEDITERRANEAN FLOUR MOTH

Adult wingspan is about one inch. The front wings are pale gray with irregular bands; hind wings are grayish white. Caterpillars are about 1-1/2 inches long and yellowish to pinkish white.

Development. Eggs are laid singly on food surfaces; a female can lay a total of 100 to 600 eggs. Hatching is in about five days. Caterpillar development is completed in 10 weeks. There are four or five generations per year. Full-grown caterpillars may be found far from the infested site.

Habits. Foods that may be infested by this moth include flour, nuts, seeds, beans, dried fruits, flour, and chocolate. The adult moths can be collected in UV light traps.

Mediterranean flour moth

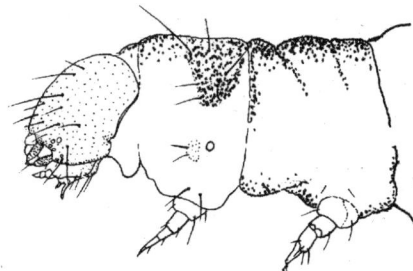
Flour moth caterpillar

CLOTHES MOTHS

The wingspan is about one inch. They are weak flyers and do not move far from the site of caterpillar feeding. Adults rarely fly to lights and are not active in lighted areas. Caterpillars feed on animal horns and woolen fabrics.

WEBBING CLOTHES MOTH

Adult wingspan is about 1/3 inch, and the body is dark yellow to reddish brown. Wings are uniformly gray, without dark spots. Caterpillars are yellowish white.

Development. Eggs are deposited singly or in batches of about 25 between threads on cloth surfaces. The female deposits 40 to 100 eggs in her lifetime. Hatching occurs in about seven days. Caterpillar development includes five to 45 stages (molts), and can last from 35 days to 2-1/2 years. Caterpillars make cocoons for pupation. Males live 13 to 79 days; females 10 to 48 days.

Habits. Clothes moth caterpillars feed on wool clothes, natural carpets, furs, stored wool, and piano felts. This clothes moth usually stays in the immediate area of the infestation. It tends to flutter about rather than fly in a direct, steady manner.

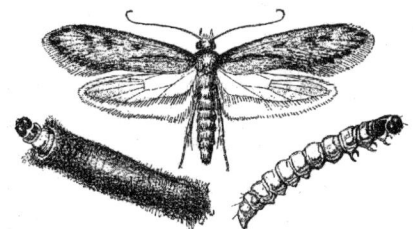
Clothes moths and larvae

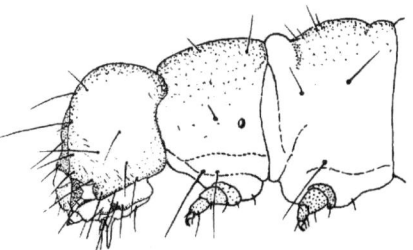
Webbing clothes moth caterpillar

CASEMAKING CLOTHES MOTH

Adult body is grayish yellow. Front wings have three dark spots on the middle; the hind wings are yellowish brown without spots. Males are active flyers, but female moths are slow and relatively inactive. Caterpillars are yellowish white.

Development. Eggs are laid singly or in small groups; females deposit 37 to 48 eggs. Hatching occurs in four to seven days. Caterpillar development is completed in 68 to 87 days, and pupation takes place in the larval feeding case after both ends are sealed. There are three or four generations per year.

Habits. The casemaking clothes moth also will feed on stored plant materials, such as spices and tobacco.

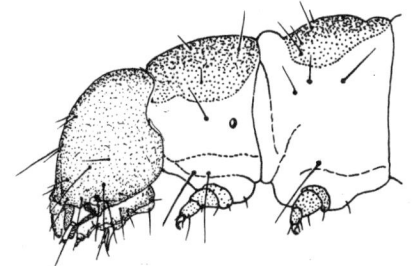
Casemaking clothes moth caterpillar

Plaster bagworm adult

Plaster bagworm case

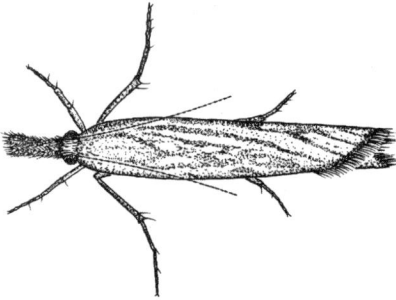

Sod webworm moth

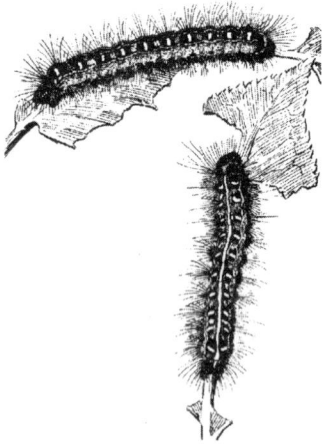

Tent caterpillars

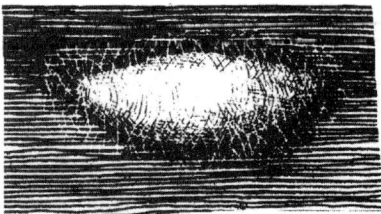

Tent caterpillar coccoon

PLASTER BAGWORM

Adult female wingspan is about 1/2 inch; the body is gray with four spots on the front wings. This caterpillar is yellowish white, with dark plates on the segments behind its head; it remains in a case with only its head protruding. The case is about 1/2 inch long.

Development. Eggs are deposited singly or in batches, hatching occurs in about 10 days. Caterpillar development is completed in about 50 days and there are seven stages before the pupal stage. The life cycle from egg to adult moth takes 62 to 86 days.

Habits. Adult moths are capable of long flight and will rest on walls or edges of the floor. The caterpillar moves by crawling on a substrate and pulling the case behind it. Cases can be found on wool rugs and wool carpets, or hanging on curtains, under buildings, from sub-flooring, and joists. The plaster bagworm requires high humidity to survive, which limits its range to southern and coastal regions.

SOD WEBWORM

This moth is gray and about one inch long. It has long palps extending from the head and long antennae extending back over its body. Its wings usually are wrapped around its body to create a tubular appearance; this distinguishes it from other moths that fly to outdoor lights.

Development. Eggs are deposited singly or in batches in turfgrass; the adults typically move in short flights over the grass. The caterpillars feed on blades of grass; the pupa is formed in the thatch layer of turf. Some species have one generation per year, while others have two or three generations.

Habits. The adults are active at dusk and will fly for about an hour after sunset. They are strongly attracted to lights at night and will come to outdoor lights and to windows in lighted rooms. There are numerous species, some occur in spring, some in summer, and some in fall. The moths may be captured in UV light traps positioned indoors, especially those near doors and windows. They can be mistaken for flour moths.

Tent caterpillar damage

EASTERN TENT CATERPILLAR

Adult body is yellowish to dark brown; the forewings have two yellowish white lines. Caterpillars are about two inches long, with black head and scattered long hairs. There is a white stripe, bordered with reddish brown and black lines running the length of its body.

Development. Eggs are laid in masses of 150 to 250 that encircle small twigs on host trees. Hatching occurs the following spring, about the time new leaves appear on the tree. Caterpillar development is about three weeks; full-grown caterpillars leave the tent and wander in search of a place to pupate. Pupation occurs in silken cocoons, covered with a yellowish powder. There is one generation per year.

Habits. The tent is constructed in a crotch of a tree, and it is expanded with the growth of the caterpillars. Preferred hosts are wild cherry and apple trees, but it also attacks shade and fruit trees. The caterpillar usually leaves the tree to spin its cocoon, which often is formed on the sides and eaves of houses. The oval shape and yellowish coloring of the cocoon are characteristic of this species.

BAGWORM

Full-grown caterpillars are about 1-1/2 inches long and dark yellow to light brown. Caterpillars attack evergreen and deciduous trees, with the most commonly infested trees being firs, juniper, pines, spruce, maple, sweet gum, and sycamore.

Development. Eggs are laid inside the bag and hatch the following spring. First-stage caterpillars leave the bag and move onto leaves to feed and form their own bags. Caterpillar development is completed in fall. Caterpillars leave their host trees or shrubs to pupate; they pupate in the case.

Habits. Males leave the case after development, but the wingless females remain in the case. In fall, females extend their abdomens to the outside of the bag to attract males for mating.

Bagworm

CHAPTER 11

TERMITES

Termites have successfully invaded nearly all habitats and can use dry, wet, above-ground, and below-ground wood for food and nesting sites. They live in highly organized colonies, with each individual having distinct responsibilities and the body form to match its task. Workers are small, search for food, and maintain the colony; soldiers are large and defend the colony; and the queens simply lay eggs.

Termites have incomplete metamorphosis, the development stages are egg, nymph, and adult. Termite colonies have a functional male (or king) to mate with the queen in the nest. This is different than ants, for which only the female survives after the mating flight, and total egg production is limited by the single mating during that flight. In termite colonies, mating and egg production is ongoing throughout the long life of the queen. Development consists of a series of nymphal stages; these individuals develop into soldiers and workers, and they are recognized by their size and task in the colony. Swarmers are males and females, and usually are produced annually.

Dispersal of winged termites (swarmers) from colonies is an important means of establishing new colonies. Swarmers are produced by well-established colonies, and this occurs three to five years after the colony founding. The release of these winged forms is usually restricted to certain times of year, and specific times of day. Swarming is synchronized with regional and local weather conditions; it occurs during warm months in temperate regions. Soldiers and workers excavate exit holes and protect the emerging swarmers. The number of swarmers produced can be as much as 43% of the total individuals in the colony.

Drywood termite nests are not built in contact with the soil; they depend on wood moisture for suitable conditions for the colony. Infestations may go unnoticed because their feeding leaves a thin veneer of wood at the surface that hides the damage. Their rounded fecal pellets are ejected from the galleries, and piles of these pellets are a sign of infestation. Dampwood termites infest wet and decayed wood, but feeding can extend to sound wood. Subterranean termite nests are built in soil or in wood buried in soil, but the termites also forage above ground. Secondary above-ground nests are connected to the primary nest in the soil by shelter tubes.

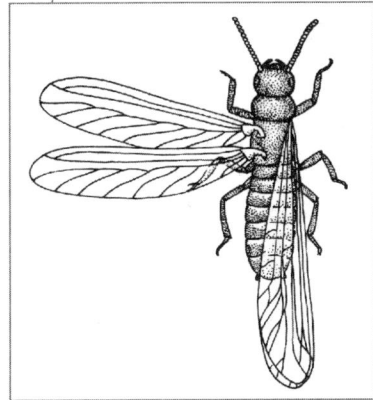

Termite—winged

Termite queen

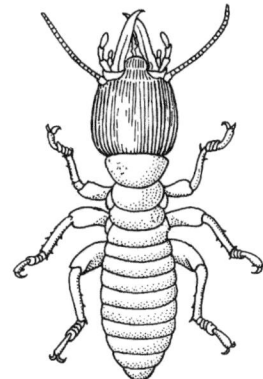

Termite soldier

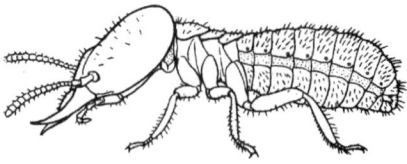
Eastern subterranean termite soldier

Eastern subterranean

Southeastern subterranean

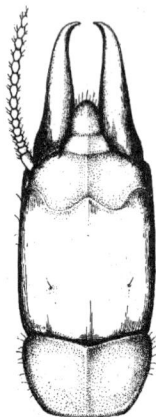
Dark southeastern subterranean

SUBTERRANEAN TERMITES

These termites nest in soil, but feed on wood above ground. After many years infesting a structure, a large portion of the colony may be located in wood above ground. The Formosan termite is a subterranean species, but the queen may be in an above-ground nest.

Colony. Subterranean termites have a colony structure that consists of a number of feeding sites. The traditional enlarged queen is not common in these colonies; egg production is primarily by supplementary queens. Workers and soldiers move between the ground and food source through earthen shelter tubes. Most colonies produce several swarms in spring, but swarms also may occur in heated building in mid-winter and continue into spring.

Galleries. Workers excavate galleries that usually follow along the natural grain of the wood. The galleries are lined with a layer of soil, which helps to maintain humidity in the wood. The galleries are below a thin layer of wood and not visible from the outside. There are no fecal pellets in the galleries of subterranean termites.

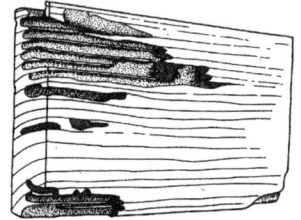
Subterranean termite wood damage

EASTERN SUBTERRANEAN TERMITE

Soldier head is longer than it is broad; the sides are nearly straight and taper to the broadly rounded posterior.

Swarming in northeastern states occurs about mid-day in April and May; in the southeast, flights occur from March to May; in Florida, these termites swarm in February. Swarming may occur at any time of year indoors.

SOUTHEASTERN SUBTERRANEAN TERMITE

Soldier head is pale yellowish brown and darker at the front; mandibles are as long as the width of the head.

Swarming is in late summer and fall, usually from July to August. Swarmers are sometimes confused with pavement ants, which are light brown and often swarm in fall.

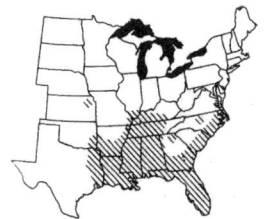
Southeastern subterranean distribution

DARK SOUTHEASTERN SUBTERRANEAN TERMITE

Soldier head is about as broad as it is long. The mandibles are long.

Swarming flights are in March and April. This is a common species in the southeastern U.S.

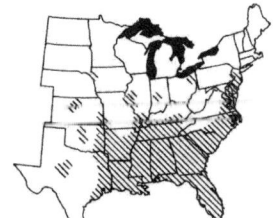
Dark southeastern subterranean distribution

WESTERN SUBTERRANEAN TERMITE

Soldier head is at least two times longer than it is broad and is yellowish brown.

Swarming flights occur during the day in November to January following rainfall; flights may also occur from February to June.

Western subterranean distribution

ARID-LAND SUBTERRANEAN TERMITE

Soldier head is brownish yellow, and the sides are nearly parallel. Mandibles are about as long as the width of the head.

Swarming flights occur in spring and fall; in the central Rocky Mountains, swarmers may emerge in April and February.

FORMOSAN TERMITE

Soldiers have an oval-shaped, light-brown head and dark-brown mandibles; the body is yellowish brown.

Swarming flights are at dusk to about midnight, from March through June. Swarmers have a strong tendency to fly toward outdoor lights at night.

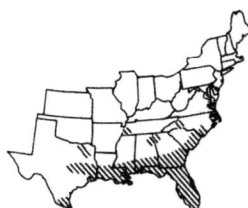

Formosan termite distribution

DRYWOOD TERMITES

These termites nest in structural wood that has 12% to 15% moisture content. The colonies do not require contact with the soil, but are able to survive in structural timbers in attics and house framing, as well as in furniture.

Colony. A large colony may contain several thousand individuals and survive for about 10 years. A typical colony may consume about a half pound of wood per year.

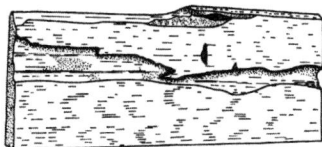

Drywood termite damage

Galleries. Galleries usually do not follow the natural grain of the wood. Dry pellets are stored in portions of the nest, or cast out through "kick-out" holes in a gallery. The pellets are six sided with distinct ridges; the ends are rounded.

SOUTHEASTERN DRYWOOD TERMITE

Soldiers are yellowish brown. The anterior margin of the pronotum is deeply concave at the midline.

Swarming flights occur in early evening after sunset in May and June. Individuals in a swarming flight may fly to lights.

Habits. Damage is to the woodwork in buildings and hardwood furniture.

Western subterranean

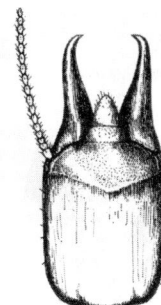

Arid-land subterranean

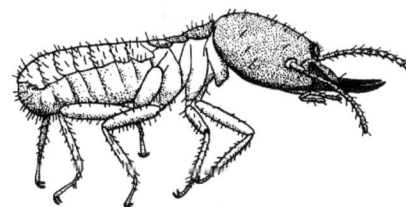

Formosan soldier

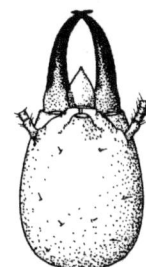

Formosan soldier head

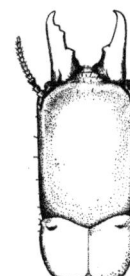

Southeastern drywood soldier head

WESTERN DRYWOOD TERMITE

Soldier head is reddish brown; mandibles are black. The third segment of each antenna is enlarged.

Swarming. Low-population flights occur from April through November on warm sunny days.

Habits. They infest rafters, ridgepoles, and sheeting in attics. In living areas, they infest window frames and sills, doorframes, and floor joists. They also infest wooden furniture. Signs of a colony are piles of brown fecal pellets below small holes in the infested wood, particularly where the outer walls of the wood are thin.

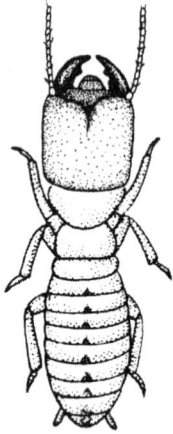

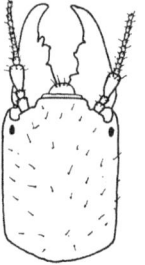

Western drywood soldier head

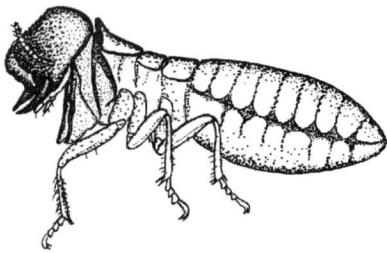

WEST INDIAN DRYWOOD TERMITE

Soldier head is brown to reddish brown.

Swarming flights occur in May and June. Damage is usually to floors, woodwork, furniture, and small wooden objects.

West Indian drywood Termite distribution

Habits. Fecal pellets are small, round, and dry; they usually are expelled from the galleries and collect in piles below infested wood.

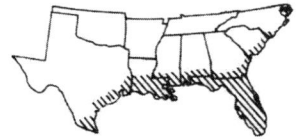

DAMPWOOD TERMITES

These termites nest in wood with a high moisture content, such as rotten or decayed wood; they do not require contact with the soil. They excavate large galleries in wood, and fecal pellets are scattered in the galleries or discarded through small holes. The pellets are six sided, but lack the distinct ridges of drywood-termite pellets.

Colony. It may take several years for a colony to have 4,000 workers. The maximum size colony for some species is only about 1,500 individuals. Swarming occurs in late summer and fall after rainfall; they are attracted to outdoor lights.

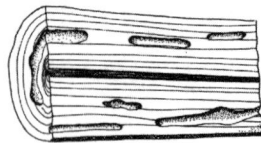

Galleries. The galleries do not follow the natural grain of wood; they are not lined with soil. They often contain accumulations of frass pellets. There are small holes for the fecal pellets to be expelled from the galleries.

Dampwood termite damage

FLORIDA DAMPWOOD TERMITE

Soldier is yellowish brown; the head is longer than it is broad; mandibles are reddish brown and as long as the width of the head. Antennae are as long as the head.

Swarming flights occur at dusk from October through January. Swarmers are attracted to lights at night.

Florida dampwood termite distribution

Habits. Natural sites are damp and dry wood in logs of tidal mangrove swamps, or in pine woods near seacoasts. This termite infests wood around building foundations.

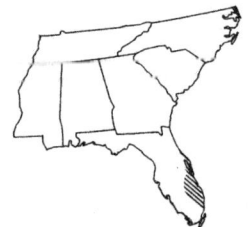

Western drywood soldier

West Indian drywood soldier

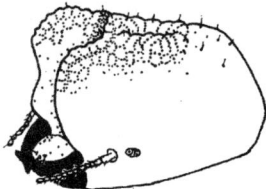

West Indian drywood soldier head

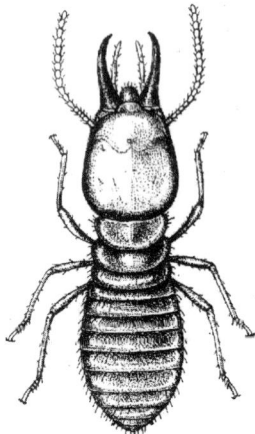

Florida dampwood soldier

EASTERN DAMPWOOD TERMITE

Soldier head is light brown; the mandibles are blackish brown, with brown at the base.

Eastern dampwood termite distribution

Swarming flights occur in March, June, October, and November.

Habits. Natural sites include the decayed wood of trees, tree stumps, logs, and branches. This termite also will infest living trees, including those that are grown as indoor plants in large buildings and shopping malls.

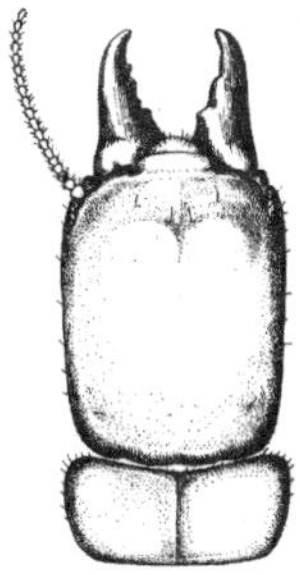

Eastern dampwood soldier head

PACIFIC DAMPWOOD TERMITE

Soldier head is longer than it is broad, and dark red to reddish black; the mandibles are black. The sides of the head are concave, and the head is somewhat narrower in front than behind. Fecal pellets are rounded and usually the color of the infested wood.

Pacific dampwood Termite distribution

Swarming is at dusk, usually before sunset, and occurs in May to November. Swarmers are attracted to lights at night.

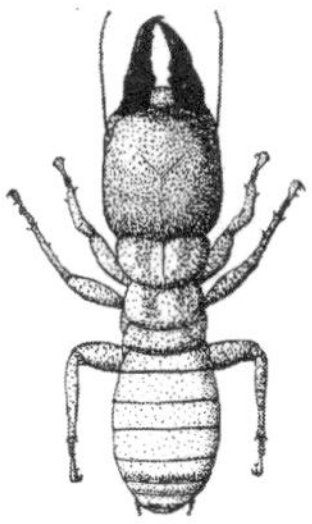

Pacific dampwood soldier

NEVADA DAMPWOOD TERMITE

Soldier head is not narrow at the front, and the sides are nearly parallel.

Navada dampwood termite distribution

Swarming occurs at dusk, and flights have been reported in January, July, August, and September. Swarmers are attracted to lights at night.

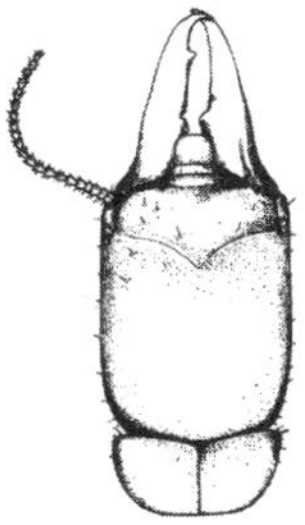

Nevada dampwood soldier head

_____

_____

_____

_____

_____

_____

_____

_____

_____

_____

CHAPTER 12

TICKS, MITES, SOWBUGS, AND PILLBUGS

Ticks are blood-sucking parasites of mammals, birds, reptiles, and amphibians. They are closely related to mites, but are distinguished by their body shape and large size. The head, thorax, and abdomen in ticks are fused into a single unit. Males are typically small and may be unnoticed, but females often have an enlarged abdomen that is filled with a recent blood meal or with thousands of eggs. Ticks have seasonal abundance: nymphs and adults actively search for a host in early spring; females usually lay eggs in summer; adults search for a host in fall; and most early-stage nymphs overwinter, and then search for hosts and a blood meal in spring.

The mouthparts of ticks are formed into an elongated organ that projects forward from the front of the head. This is the portion that is inserted into the skin of the host animal. It has rows of backward-directed spines that hold the tick in place and make it difficult to remove a tick once it starts to feed. Often, this portion of the mouthpart remains in the skin when a tick is removed, but there are no ill effects from this.

Mites are generally oval with no distinct body regions. They are abundant in soil, and many are parasitic on insects and other animals. Others are scavengers on plant and animal matter, and some feed on live plants. There are several species of mites that live in bird nests and feed on the blood of adult and nestling birds. When these nests are abandoned in late spring, or when the nests are disturbed, the mites will move away. Bird nests may be built on window ledges, in the external openings of clothes-dryer vents, or in other locations on the house. When these nests are abandoned, mites may move indoors. They cannot feed on humans, but they can give a mild bite on the skin.

Sowbugs and pillbugs are closely related to lobsters, crayfish, crabs, and shrimps. These animals usually remain on or in damp soil or other moist habitats, and are active at night when humidity is high. They often gather together to reduce body evaporation and maintain water balance. Pillbugs resemble sowbugs, but differ in their body shape and behavior. The pillbug's abdomen is rounded at the end, but in sowbugs there is a pair of pointed tails at the end of the abdomen. When a pillbug is disturbed, it bends its body head to tail to form a compact ball; sowbugs are not capable of forming a compact ball.

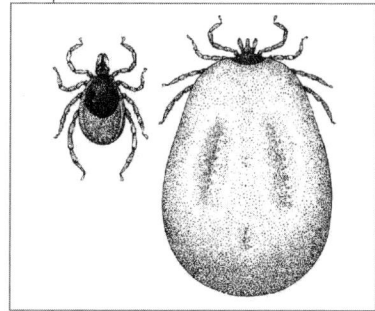

Deer ticks

Tick mouthparts

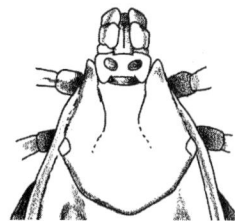

Tick head

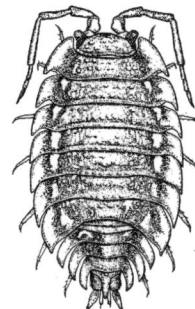

Sowbug

Deer mouse

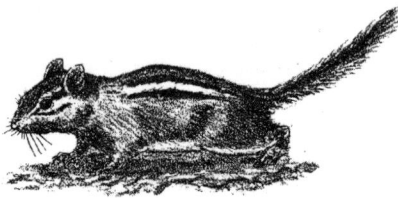

Chipmunk

TICKS

The abundance of deer mice, chipmunks, skunks, raccoons, and white-tail deer in suburban and urban areas has spread ticks to these habitats. Ticks take a blood meal from a variety of animals, including dogs, cats, and people. Their small size and undetectable bite makes them difficult to prevent and control.

Life history. During its development, a tick will take blood meals from three distinct hosts. Eggs are laid on the ground; the female may

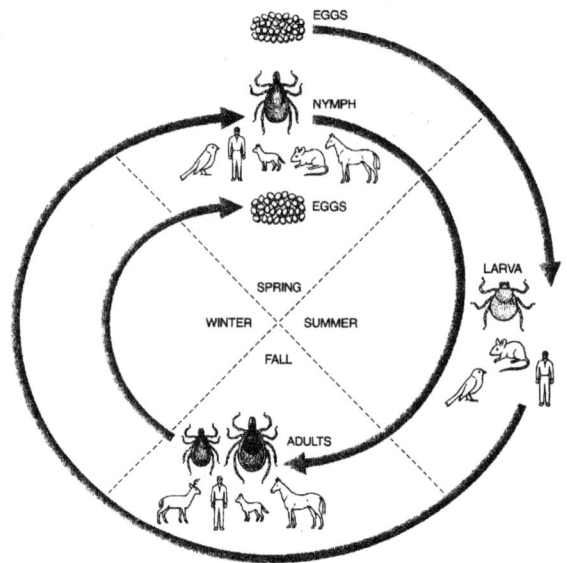

Lyme disease—seasonal lifecycle of ticks

deposit thousands of eggs at one time. After hatching, the larva (called a seed tick) finds and feeds on its first warm-blooded host. Soon after feeding, the tick drops to the ground and molts to the next stage. Then the nymph feeds on a second host, and drops off to molt; the adult feeds and mates on a third host, and drops off. The female lays eggs soon after she leaves the last host. Each stage remains on the host for one to three days and takes one blood meal.

Disease. The blood-feeding habits of ticks and their use of multiple hosts enable the spread of disease to domestic animals and humans. Tick-borne diseases include Rocky Mountain spotted fever and Lyme disease. Several species of ticks occur in urban and suburban areas and use domestic and wild animals as hosts. Rocky Mountain spotted fever is widely distributed, and in spite of its name, this disease is prevalent in the eastern United States.

Lyme disease is the most common illness carried by ticks in the U.S. Lyme disease affects the joints, heart, and nervous system of people suffering from the bacteria. It is difficult to diagnose. This disease is transmitted to people when they are bitten by the nymph stage of deer ticks. The primary sources of the disease for the ticks are deer mice and chipmunks in urban and suburban habitats. The mice and chipmunks are infected but not harmed by the bacteria. The immature stages of deer tick that feed on these small animals become infected, and transmit the bacteria to people (and dogs or cats) when they are biting and taking a blood meal. Immature deer ticks are very small and difficult to see, so they easily can go undetected. Adult deer ticks do not transmit the disease.

Deer mice and chipmunk populations are difficult to control; the best strategy for reducing the potential of getting Lyme disease is control of the immature stages of the deer tick. That can be done easily and effectively by treating the mice and chipmunks in the area with a topical insecticide, such as Tick Box Technology, that delivers insecticide to kill the ticks on small rodents when they enter a ground-secured box and rub against a wick treated with a low concentration of insecticide. This method reduces the number of immature ticks in an area, and significantly reduces Lyme disease.

LONE STAR TICK

Adult males are uniformly light brown to brown. Females have a pale white spot at the middle of their backs. This white spot is the origin of its common name.

Habits. Immature stages crawl to the top of grass stems and other vegetation, and from there, attach to host animals. Adults and immatures overwinter in soil and leaf litter.

AMERICAN DOG TICK

Adult males have an irregular pattern of white marks on their backs. These marking are easily seen on the males, but the females have a swollen abdomen and the white markings are hidden. The abdomen of the female may be grayish blue, and sometimes slightly green.

Habits. Immatures feed on mice and voles. Adults feed on dogs and other large animals, including humans. Female ticks engorge in six to 13 days. Unfed adults live about two years. This tick is most common in spring and fall, but it also attaches to and feeds on animals in winter.

BROWN DOG TICK

The male is uniformly reddish brown; engorged female is grayish blue to light green.

Habits. Immatures crawl on indoor walls and attach to pets and people; they can survive about eight months without food. Nymphs feed for about six days then drop off the host and molt to adults in 12 to 29 days. Adults attach to dogs or other animals, and suck blood for six to 50 days. Development from egg to adult is completed in about 60 days.

DEER TICK

Body is uniformly brown and has a dark brown to black dorsal plate. The dorsal plate of the male nearly covers the abdomen, but it is small in the female. This species feeds on birds, small and large mammals, and humans.

Immature stages commonly feed on deer mice, which is the primary reservoir for Lyme disease in northeastern U.S. Lyme disease is transmitted from mouse to mouse and mouse to humans by the immature stages of deer ticks.

Adults attach to white-tail deer for overwintering. The adults do not move from host to host. Adult ticks are not responsible for transmission of Lyme disease.

MITES

A mite's body generally is oval with little differentiation of the two body regions. Mites are abundant in soil and water (fresh and salt), and many are parasitic on insects. There are numerous species that are associated with stored food products or flour. Clover mites live and feed on outdoor plants, but population explosions can make them indoor invaders. A chigger is the immature stage of a mite; it can cause skin irritations when it attacks humans.

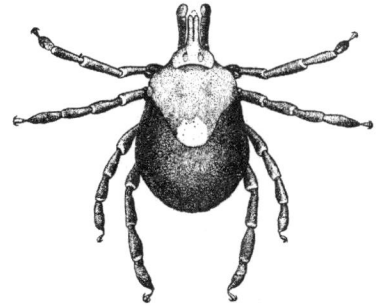
Lone star tick

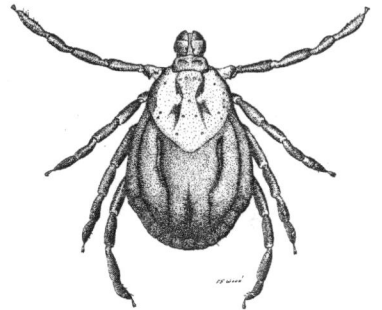
American dog tick

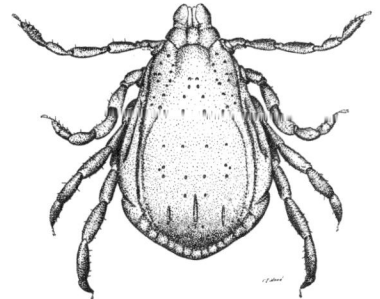
Brown dog tick

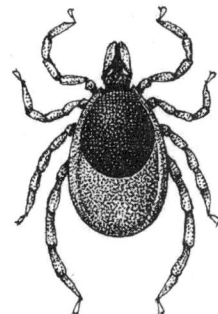
Deer tick nymph

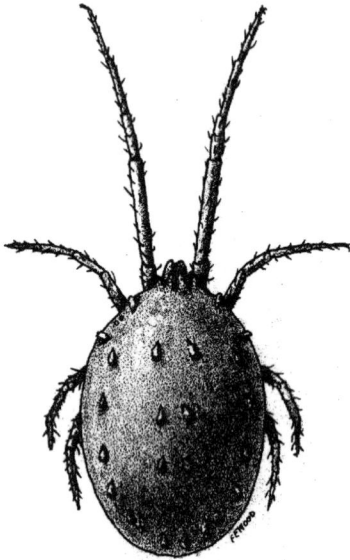

Clover mite

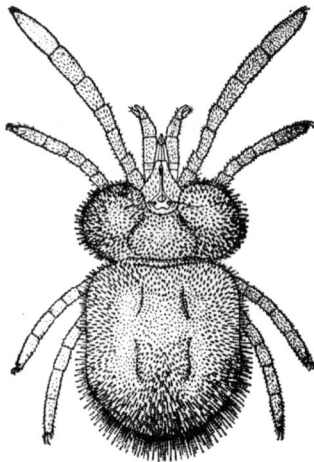

Chigger adult

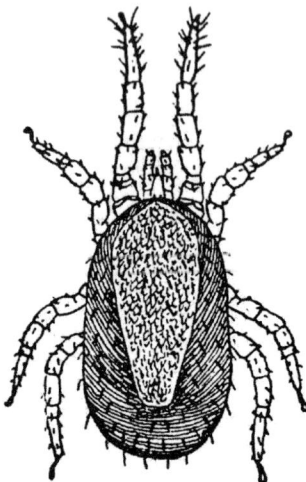

Bird mite

CLOVER MITE

Body color is reddish brown to dark green; front legs are longer than the body. The female lays about 70 bright red eggs in her lifetime. Development from egg to adult takes about 30 days. Eggs laid in late fall hatch the following spring.

Immature stages are typically bright red. They are active in turfgrass and low vegetation where they feed on plant sap.

Adults are active in eastern U.S. from October until May. They sometimes occur in large numbers on the sunny sides of buildings.

Habits. These mites climb on the outside of buildings and enter through windows and doors. Clover mites often occur in new lawns or recently established turfgrass.

CHIGGER, REDBUG

Body of the adult is bright red (redbug), and has a velvety appearance. Females lay about seven eggs per day in soil; hatching occurs in about six days. Development from egg to adult takes about 55 days.

Immature stages do not burrow into the skin (contrary to popular belief), but they attach to the base of a hair. They slowly move into the skin along the base of the hair follicle. Chiggers do not suck blood, but they inject saliva that breaks down cells. The action of this digestive fluid causes irritation and itching.

Adults are free living (not parasitic); they overwinter in the soil. There are one or two generations per year.

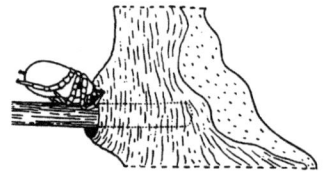

Chigger feeding

Habits. During the summer, people walking in tall grass and low vegetation may be bitten by chiggers and suffer itching for several hours to several days.

BIRD MITES

Adults are just barely visible to the naked eye. Unless they move, they are very difficult to see. Their color is translucent white; after a blood meal, they are reddish brown.

Immature stages can complete development in about 12 days. This short life cycle results in large populations in bird nests during spring when they attack young birds.

Habits. When the number of mites in a nest becomes too great, or when the young birds leave the nest, the mites will migrate away from the nest. This migration can result in mites entering buildings, especially when nests are located near windows or vents. Bird mite infestations occur in late spring to early summer when bird nests are being abandoned by the young birds.

Bites. Indoors, bird mites may crawl onto the skin and try to bite humans, but they cannot break the skin to feed. However, people may develop a rash or itching at the site of the bite.

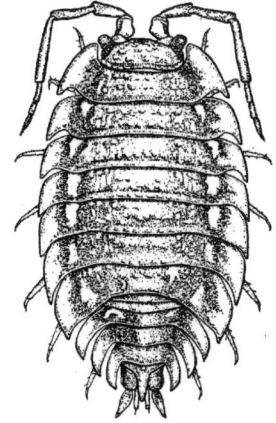

Sowbug

IMAGINARY MITES

Delusory parasitosis is the condition in which individuals believe they are being bitten by something, and it is usually thought to be mites. Typically, a person will feel something crawling on his or her skin and believe he/she is being stung or bitten. The cause is often described as something very small, black and white, and sometimes V shaped. Frequently these individuals use tape to try to capture or remove the "mites" from their skin.

Habits. People with delusory parasitosis believe the creatures are living or hiding in a variety of household materials, including clothes, bedding, furniture, and other items. Usually the problem is only at home, but it can also occur a workplace. Doctors, including dermatologists and psychiatrists, regard this condition (Morgellons), the belief that there is a pathogenic infestation in spite of no medical evidence, as a form of delusional infestation.

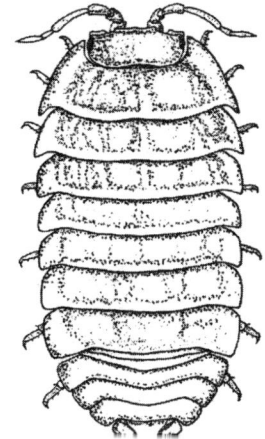

Pillbug

SOWBUGS, PILLBUGS

Body is brown and uniformly smooth and glossy with some gray to dark-gray patches and two pale longitudinal lines. They have well-developed eyes. Eggs are retained in a brood pouch; hatching occurs in about 50 days. The number of young in each brood ranges from 24 to 88.

Immature stages and adults feed at night on decaying plant material, but they will attack tender plants.

Adults live about two years; there are one to three generations per year.

Identification. The sowbug's abdomen ends with two pointed tails; the pillbug's is rounded, without tail-like projections. Pillbugs roll into a compact ball when disturbed, sowbugs do not.

_____

_____

_____

_____

_____

CHAPTER 13

SPIDERS, CENTIPEDES, MILLIPEDES, AND SCORPIONS

Spiders have two body regions: the cephalothorax and abdomen (insects have three regions: head, thorax, and abdomen). Spiders breathe through two pairs of lungs that open on the underside of their abdomens. The abdomen has specialized glands and structures for making and manipulating silk.

Spiders are predators of insects, sowbugs, and even other spiders. Some hunt during the day, some at night; some spiders specialize in capturing crawling insects, others in those that fly. Most spiders can deliver a poisonous bite, but few have mandibles that can penetrate human skin; when they do, the venom is usually harmless. Spider bites are similar to mosquito bites: there is a small swelling and itching for a short time.

A characteristic feature of spiders is their ability to produce silken threads by glands in the abdomen. Spider silk is a protein material with the strength of nylon, but it can be stretched by 31%, compared with only 16% for nylon. The silk is used to construct a snare to catch flying insects. The spider remains in or close to the elaborate web to quickly capture and feed on the prey. The most recognized webs are those made by the orb weaver spiders; these can measure several feet in diameter.

Centipedes have one leg per segment, and the number of legs ranges from 13 to 181. The first body segment behind the head contains poison claws that are used to capture prey. These arthropods are nocturnal and occur in moist habitats. Most centipedes are predators of insects. The house centipede is the most common indoor centipede. It is fast moving and can climb walls and ceilings; it is a predator of spiders and small insects. These centipedes have 15 pairs of very long legs. The long antennae move in a whip-like manner over the body.

Millipedes have two pairs of legs on most of their body segments, and they have numerous body segments. They range in color from reddish orange to dark brown and black. Millipedes typically occur in moist or wet habitats. Many species curl up or form a compact spiral when disturbed. Food for millipedes is a variety of decomposing plant and animal material.

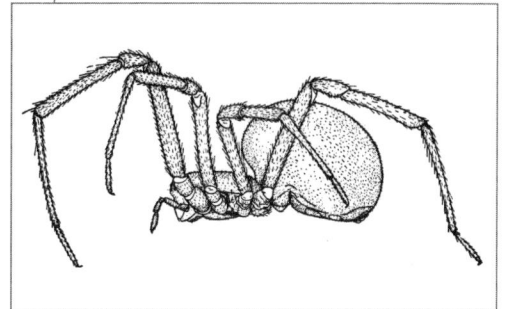

Spider

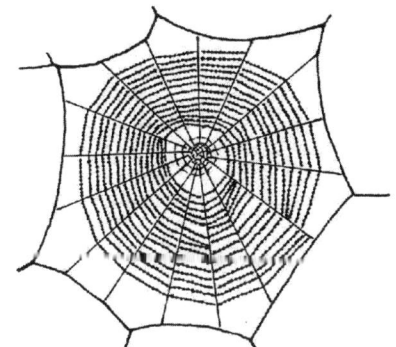

Spider web

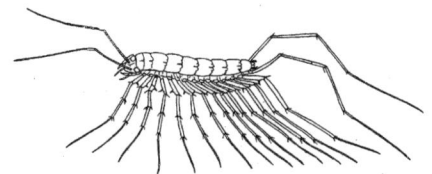

House centipede

Millipede

AMERICAN HOUSE SPIDER

The cephalothorax of the American house spider is yellowish brown; the abdomen is grayish white to brown. Legs of the male are orange; female legs are yellow with brown bands at the ends of the segments.

Egg sacs are brown, oval, or pear-shaped and usually placed in the web. Females may produce as many as 17 egg sacs, with a total of 3,794 eggs per lifetime. Adults are present year round and some individuals live for two years.

Habits. This spider frequently occurs in outbuildings and in houses. It makes webs in corners of rooms and frequently in the angles of windows. The webs are easily identified by the long strands of silk that connect a complex web to the surface below.

American house spider web

DOMESTIC HOUSE SPIDER

Males are pale yellow, with two gray stripes; the abdomen has irregular gray marks. Legs are long and distinctly banded.

Egg sacs are produced throughout the warm season and hatch in about 39 days. Adults live for several years, and males and females usually occur together on the same web. Females usually remain at the web, but males wander in the house, searching for food.

Habits. Natural habitats are under stones and in rock crevices. Indoors, they occur in cellars, and dark corners of rooms; they also can be common in outbuildings.

HOBO SPIDER

Males and females have a leg span of about 1-1/2 inches. Legs are brown without bands. Abdomen is brown and white, with distinct pale chevrons along the center.

Egg sacs are produced in fall, and the females often remain with them through winter; eggs hatch in spring.

Bites often occur without provocation, and because of this behavior, it is called the aggressive house spider.

Habits. These spiders occur along rock walls, firewood piles, and house foundations. Males and females enter houses during fall and winter after exposure to cold. Although these spiders have been blamed for causing skin damage similar to brown recluse spider bites, there is no evidence that this is true.

Hobo spider distrubution

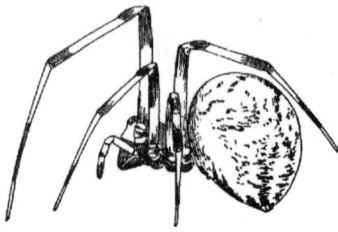

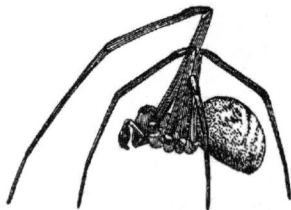

American house spider
(above: female, below: male)

Domestic house spider

Hobo spider

CELLAR SPIDER

The bodies of both the male and female are pale yellow except for a gray mark in the center of the cephalothorax. Abdomen is elongated, more than twice as long as it is wide. The legs of these spiders are extremely long and thin.

Habits. The webs are sheet-like and not easily seen. The egg case is usually carried by the female until it is ready to hatch. This is the most common spider throughout the continental U.S.

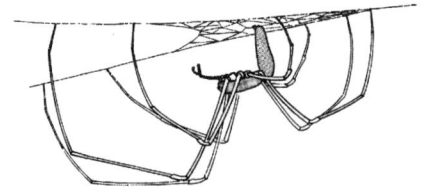

Cellar spider

YELLOW SAC SPIDERS

There are two common species, one is light green and the other is yellowish white.

Egg sacs are white and papery and are usually attached to the underside of objects. Females remain with the eggs until hatching. The presence of a female with an egg sac is the origin of its name. It is most common indoors in fall and spring.

Bite hurts much like a bee or wasp sting. Following the bite, there is redness of the skin, itching, and slight swelling. These conditions subside in a few days. Although these spiders have been blamed for causing skin damage similar to brown recluse spider bites, there is no evidence that this is true.

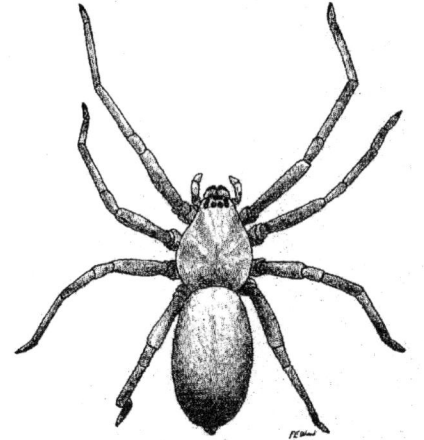

Yellow sac spider distribution

Habits. Inside houses they are found on walls and in corners close to the ceiling; they drop from ceilings on silk threads.

Yellow sac spider

WOODLOUSE SPIDER

The cephalothorax and legs are reddish orange to brown; the abdomen is cream white. The body has few setae. The fangs are shiny and project forward.

Eggs sacs are light and nearly transparent. These spiders live in a flattened, oval retreat; they hunt their prey from the retreat.

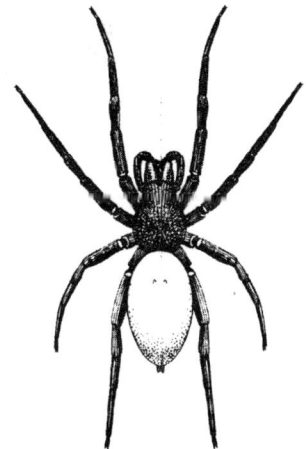

Woodlouse spider distribution

Habits. They prey on sowbugs. Natural habitats are under stones and in rock crevices; they prefer dark and humid habitats. They can be numerous indoors, particularly along baseboards in ground-level rooms.

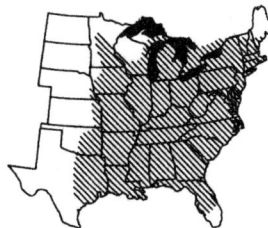

Woodlouse spider

ORB WEAVER SPIDERS

These are large, brightly colored spiders that build large orb-shaped webs outdoors. Most of the species construct webs in the shape of an orb; some females build hiding places that are separate from the large web. The most common orb weavers remain in the center of the web, hanging head downward waiting for prey. The webs are often located on shrubs around houses and close to outdoor lights. Some species only build webs at night and take them down during the day.

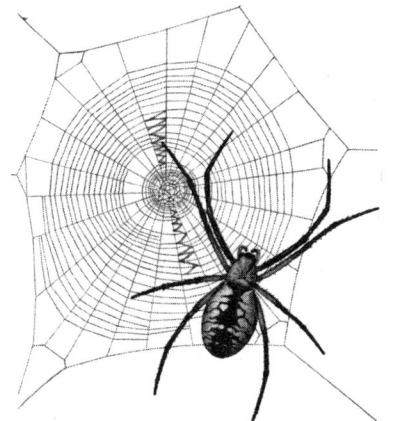

Orb spider and web

Black and yellow garden spider

BLACK AND YELLOW GARDEN SPIDER

The female's abdomen is marked with black and bright yellow or orange; it is slightly pointed at the end and curved at the sides to form a hump on each side. Front legs are entirely black; others legs have reddish brown or yellow markings with the other segments being black.

Egg sacs are light brown spheres and are placed at the edge of the web.

Habits. Webs often are constructed near outdoor lights in locations such as porches or garden furniture. Webs often contain a zigzag of thick silk, which extends above and below the center of the web.

DADDY LONGLEGS

Body is composed of one segment; the cephalothorax is joined with the small abdomen. The body is suspended above the long legs.

Habits. They drink frequently and must have water available. They are nocturnal but can be active on cloudy days and dark areas. They eat insects, especially aphids and mites. They often are misidentified as poisonous spiders, but they are not actually spiders and cannot bite.

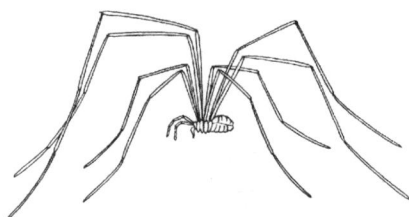

Daddy longlegs

WOLF SPIDERS

The colors and markings of these spiders are variable. Generally, they have large bodies and long legs. The species that occur indoors are usually "hairy"; some have markings that resemble the brown recluse spider.

Egg sacs will not be found because the female carries the egg sac until it hatches. The small spiders may cling to the female for a few days before dispersing.

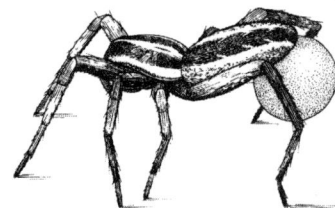

Wolf spider with egg sac

Habits. These spiders do not build webs to catch their prey or to hold the egg sacs; they move around and hunt for their food. They are called wolf spiders because of their wandering and hunting behavior. Most species are active at night. They are common indoors in spring and fall. In spring, they are searching for mates, and in fall they are retreating from cold temperatures. Wolf spiders are not aggressive and usually hunt and hide under objects. They typically remain at floor level; they are not good climbers.

Wolf spider

CAROLINA WOLF SPIDER

The body is uniformly dark brown without distinct marking. The underside of the body is darker than the top and almost black, with white markings at the base of the legs.

This is one of the largest spiders in North America. It occurs indoors in humid habitats, such as bathrooms and basements.

Wolf spider

BROWN RECLUSE SPIDER

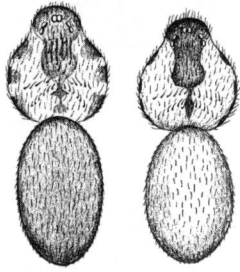

Male and female bodies are brown, except for a violin-shaped mark in the middle of the cephalothorax, with the neck of the violin directed backward. Legs are long, up to twice the body length.

Egg sacs are produced in May to July, with few in August and none in the cold months. A female produces about five egg sacs in her lifetime, with about 51 eggs in each. Adult males live 301 to 796 days and females 356 to 894 days.

Female (left) and male
brown recluse

Bites from these spiders produce mild to severe pain within two to eight hours. At the site of the bite, an open ulcer develops in one to two weeks and persists for two to three weeks. The bite from a recluse spider may result in a large wound and a lasting scar.

Habits. They occur indoors and outdoors around houses, sheds, and outbuildings.

Brown recluse spider

JUMPING SPIDERS

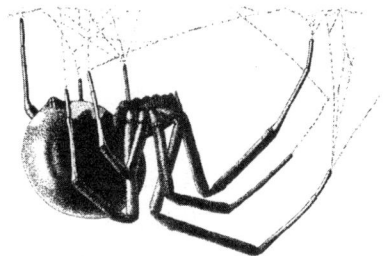

Female cephalothorax is black anteriorly and brown posteriorly, and has a pale white median stripe; male cephalothorax has lateral white stripes. Abdomen is black and the middle and lateral stripes are white.

Habits. This species occurs primarily indoors; outdoor populations are not common. It feeds on a variety of arthropods, including the German cockroach.

BLACK WIDOW SPIDER

Females are shiny black, and the abdomen is rounded; typically there is a red double triangle/hourglass mark, or a similar red mark, on the underside.

Egg sacs are globular, and placed in the web. Females produce about 10 egg sacs; total egg production may exceed 2,500. Immature spiders overwinter and become adults the following year. Males live 28 to 40 days, and females live one to two years.

Bite is painful at the site; other effects include increased heartbeat, raised blood pressure, and paralysis of the diaphragm muscles, which results in difficulty breathing.

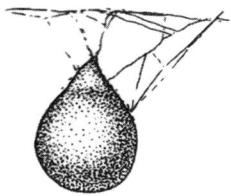

Jumping spider

Habits. The female hangs in an inverted position with legs extended, and does not move far from her web. It occurs in downspouts, firewood piles, and discarded household materials, and near vents and doors in crawlspaces.

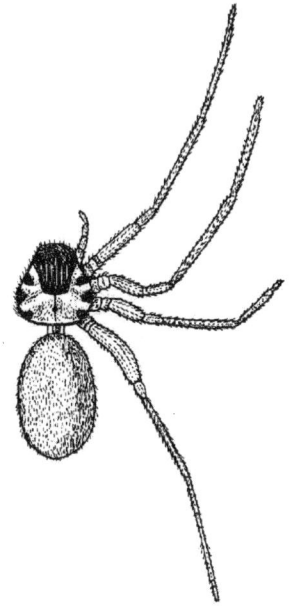

Black widow egg sac   Black widow abdomen

Black widow

Brown widow spider

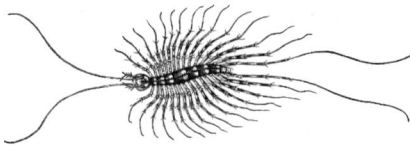

Brown widow spider

House centipede

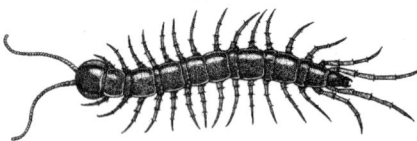

Centipede

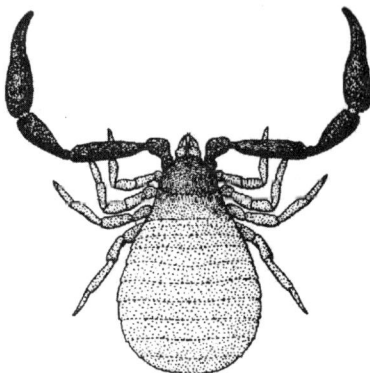

Pseudoscorpion

BROWN WIDOW SPIDER

Female abdomen is rounded but not shiny. The top of the abdomen has a highly variable pattern of spots with orange centers outlined in black with a white border. The legs are banded and are dark brown at the junction of the leg segments.

Egg sacs are distinguished by the presence of little papules on the surface.

Brown widow egg sac

Bite often is considered more toxic than the black widow, but there seems to be no evidence for this.

Habits. Webs are small and placed in well-lit areas, especially those that are lighted at night.

HOUSE CENTIPEDE

Adults are about two inches long; the body is grayish yellow with three longitudinal dorsal stripes. The antennae and 15 pairs of legs are very long; the legs are banded with white.

Eggs are placed into crevices; hatching occurs in 30 days. Development is completed in about 40 days; adults can live for several years.

Habits. This centipede feeds on house flies, cockroaches, moths, and spiders. The long legs and rapid movement on walls and ceilings make this centipede a worrisome pest indoors.

COMMON CENTIPEDE

Adults are about two inches long. The legs, antennae, and plates on the body are uniformly blue or gray with a blue tint (Virginia, North Carolina); dull gray and green (Florida); yellowish brown with blue or gray bands (central and western Texas); or uniformly blue (eastern Texas).

There is one pair of legs per segment. They are capable of rapid movement in leaf litter and soil and will quickly tunnel into soil.

Habits. This centipede occurs in mulch and other organic ground cover surrounding buildings. It can move indoors through openings around doors and windows at ground level.

PSEUDOSCORPIONS

These arthropods are about 1/4 inch long. They resemble true scorpions because they have large claws, but their bodies are short, and they do not have stingers.

Habits. They are common in leaf litter but also occur indoors or can be carried indoors on clothing. They are predators on insects and mites. They do not bite humans. They have silk glands and usually build silk cocoons to spend the winter.

TURFGRASS MILLIPEDE

Body is brown and about 1-1/2 inches long. Eggs are deposited in the soil, and hatching occurs in about three weeks. The number of legs and the body segments increase during development. Most species live several years.

Habits. Mass migrations of millipedes occur as a result of favorable conditions and a dramatic population increase, and then large numbers leave the breeding site. Two pairs of legs per segment gives millipedes considerable forward thrust. They can penetrate rotting wood and narrow openings around doors and windows.

Millipede

SCORPIONS

Body is brown and about 1-1/2 inches long. The large pincers are used for capturing prey; at the end of the abdomen is the enlarged segment that includes the stinger.

Eggs are held internally, and the female gives birth to live young which often remain on the back of the female for about two weeks.

Habits. Scorpions do not nest but establish a territory for foraging; they may occur in groups in a harborage. Changes in the local environment or habitat may cause them to move into new locations. Outdoors, they can occur in firewood and other debris; indoors, they are usually associated with water, such as in kitchens and bathrooms. Centruroides species can climb walls.

Scorpion

LAWN SHRIMP (AMPHIPOD)

Body is about 1/2 inch long, and brown to brownish black, but it turns red when it dies. They are killed (drowned) when their habitat floods.

Habits. These are terrestrial amphipods that live in damp soil under dense vegetation. They are able to jump somewhat like a flea. They can accumulate around the perimeter of houses but rarely move inside.

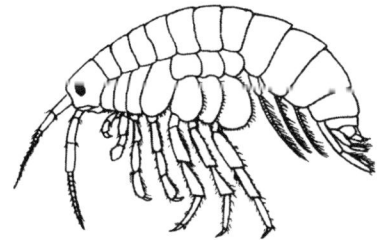

Lawn shrimp

CHAPTER 14

VERTEBRATES

Rat pair

Rats and mice are pests of residential and commercial buildings. The most common rats are the Norway rat and the roof rat. The house mouse is more widely distributed than the two rat species and more common as an indoor pest. Deer mice invade houses in fall, but they usually do not remain as permanent pests. The success of these rodents as pests is based on their ability to enter structures through small openings, utilize nearly all human food to live and reproduce, and be very wary of attempts to control them.

Norway rats and roof rats have different habitats and preferences. Roof rats prefer warm temperatures and occur primarily in coastal regions. They range along the eastern and western coasts, through the Gulf Coast states, and in Hawaii. However, in the last 10 years, they have been moving farther inland and may occur along with Norway rats in some locations. When these two species occupy the same location, they separate themselves to limit competition for food and harborage. Roof rats will nest in above-ground locations and Norway rats at ground level, typically in burrows.

Flying bat

The little brown bat and big brown bat are common in urban and suburban areas. While they are disliked or even feared by some people, they have an important role in the environment. These bats eat a large number of insects; a colony of bats can eat about 150 pounds of insects from May to September. However, their benefits often are discounted when they roost in buildings and present a health problem, because they can carry rabies. Bat control or exclusion measures are limited and closely regulated by state agencies. In general, control measures cannot be enacted during summer.

Squirrel

Squirrels often are considered cute when they are chasing each other across the grass or up trees in the yard. But these rodents can become nuisance pests in many situations. The gray squirrel is becoming dominant in many regions of the country; mild winters and the reproductive ability of these animals have increased their numbers. The most serious damage occurs when they enter roof spaces or attics by climbing the sides of houses or entering from a nearby tree. Once inside, they can damage electric wires and disrupt insulation. Their activity in attics and wall voids is an additional problem to homeowners.

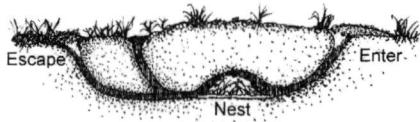

Rat burrow

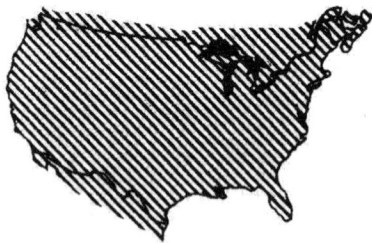

Norway rat distribution

Norway rat feces

RATS

Norway rats and roof rats feed on a wide variety of food and can nest, breed, and survive indoors and outdoors. They can gain entry to buildings through small openings they enlarge with their powerful jaws. Their teeth can cut through aluminum, lead, copper, asphalt, wood, sheetrock, plastic, and soft mortar. Their jaws can bite with about one pound of pressure,

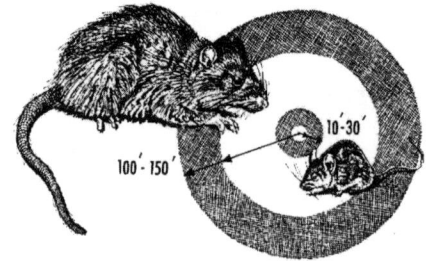

Rodent foraging range

and they take one and a half bites per second when they are intent on expanding a small opening into a passageway.

Reproduction. The number of young born increases when there is abundant food and harborage. Norway and roof rats are capable of mating and producing young at 12 weeks of age. After giving birth, they can become pregnant again in 48 hours; they will reproduce year round. Their reproductive abilities alone make these rodents a serious pest and make effective pest control critical.

Foraging territory. Rats forage soon after sunset and just before sunrise. The home range of rats and mice is generally considered to be the area regularly frequented in the search for food and water. The size of this area can vary from season to season (spring and fall), according to sex or population density. Rats have a home range of about 150 feet in diameter. However, if their food sources and shelter are secure and undisturbed, rats can live for months in an area of 60 feet in diameter. Mice have a much smaller foraging territory, generally about 30 feet in diameter.

Inspection. Signs of rodent infestation include runways regularly used between food and harborage and fresh droppings along runways and in resting sites. Urine stains along runways can be seen with UV light; rub marks along walls and entry holes are residues from body oil and dirt on the fur of rats; footprints and tail-drag marks may be seen in dusty locations or where tracking powder has been applied. Inspections must be done inside and outside, and always considering that rats climb and enter above ground level.

Rat burrows may be seen along the outside perimeter of the building. Burrows typically have at least one entryway that leads to the nest site, and a couple escape holes that may be loosely covered with soil. Chewed or gnawed entry holes usually are at the edges of doors or where the siding joins the foundation.

Norway rat

NORWAY RAT

Adult rats are about 10 inches long, and weigh about 1-1/2 pounds. Their bodies are grayish brown, but vary from grey to blackish brown to black. Their ears and eyes are small.

Food. They eat about one ounce of food and drink about one ounce of liquid per day. Their diet consists of meat, fish, vegetables, and grains.

Droppings. Their feces have blunt ends and are about 3/4 inch long; adults produce 30 to 180 droppings per day.

Habits. These rats make burrows in soil; they nest in basements and lower levels of buildings. They can carry several species of fleas, including the cat flea and the dog flea.

ROOF RAT

Adults are about eight inches long, and weigh about one pound. Body color is black to brownish gray; the underside is gray to white. Tail is hairless and about 10 inches long. Ears are large and cover the eyes if bent forward. The eyes are large.

Food. They eat about one ounce of food and drink about one ounce of liquid per day. These rats eat mostly fruits, vegetables, and grains.

Droppings. Their feces have pointed ends and are about 1/2 inch long; adults produce 30 to 140 droppings per day.

Habits. This rat usually enters and nests in upper levels of buildings; it also may nest outside in trees and dense vegetation. It burrows very little. Activity takes place soon after sunset and just before sunrise.

MICE, MOLES, AND VOLES

The house mouse is an indoor pest nearly any time of year. House mice may remain in buildings for numerous generations and have no contact with outdoor habitats. Deer mice are primarily seasonal pests; they enter buildings in fall when food outdoors becomes scarce and temperatures drop. Moles can be year-round pests, but their tunneling is most often a problem in turfgrass in spring and summer. The system of surface runways made by voles also damages turfgrass. They are active year round and can create extensive tunnels under snow-covered soil.

Reproduction. The house mouse breeds throughout the year; females have litters of three to 12 offspring five to 10 times a year. Gestation is about three weeks. Young are weaned by 21 days, and they can begin to reproduce when they are two months old. They live two to three years.

Behavior. The house mouse is a social animal; it typically lives in groups with several others. Each group occupies a territory that is indicated by scent markers; typically the members of the group have their own nests. Both deer mice and house mice urinate and defecate in their nests, and when it becomes foul in a few weeks, they move and build another nearby.

Feeding and foraging territory. House mice feed up to 20 times a day. They eat grains, fruits, vegetables, meat, and insects, but also are known to eat glue, paste, and soap. If they can find and eat food that has at least 12% protein, they can survive without drinking water every day. The home range of house mice is variable, but it is considered to be 10 to 30 feet when they infest indoors. While rats typically have a flat or one-dimensional foraging territory around their nest sites, the foraging territory of mice has to be considered in three dimensions, because they will readily climb and move in all directions in search of food.

Moles tunnel beneath the soil surface and spend little or no time on the surface. The hairy-tailed mole will tunnel beneath the soil surface during the day, but often emerges at night to feed. The exit holes are usually indicated by a mound of fresh or loose soil. The burrow includes a system of permanent tunnels that has a nest site and a large number of feeding runways close to the surface. At the surface, they find earthworms and insects for food. Moles can be active all winter, but they usually remain in the deep burrows.

Roof rat

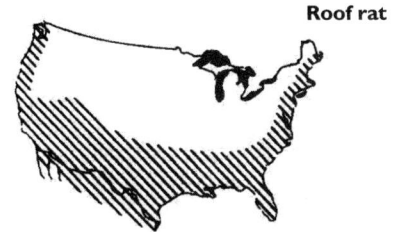
Roof rat distribution

Roof rat feces

Mouse

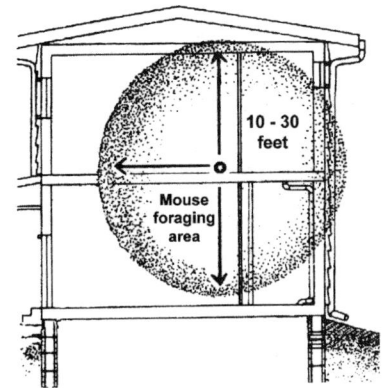
10 - 30 feet
Mouse foraging area
Mouse foraging distance

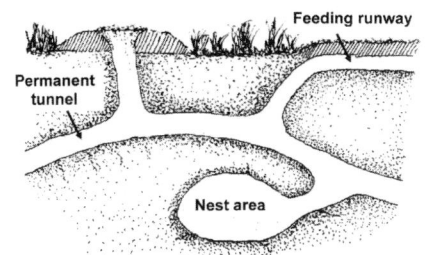
Feeding runway
Permanent tunnel
Nest area
Mole burrows

House mouse

House mouse feces

Deer mouse

Deer mouse feces

Vole adult

Vole burrows

HOUSE MOUSE

Adults are about four inches long and weigh about one ounce. Body color is typically grayish brown, and yellowish white on the underside. Tail is longer than the head and body. Ears are large, and the eyes are small and close together.

Food. These mice eat about a tenth of an ounce of food per day; water is not essential if the food contains at least 16% moisture. The food eaten includes seeds and grains. Mice visit 20 to 30 food sites while foraging, eat small amounts at a few sites, and drop feces and urine along the way.

Droppings. Their feces have rounded ends and are about 1/4 inch long; adults produce 30 to 50 droppings per day.

Habits. The house mouse is an excellent climber and can enter houses or buildings from ground level to upper stories. It can jump long distances and survive an eight-foot fall.

DEER MICE, WHITE-FOOTED MICE

Adults are about five inches long and weigh about one ounce. Body is dull orange-brown above and white below. Tail is nearly half the length of the body. Ears are large.

Food. They eat seeds, nuts, and berries; favorite foods for the white-footed mouse include black cherry pits and seeds. They commonly cache large supplies of food in wall voids, dresser drawers, cabinets, and other narrow locations. These caches can be infested with carpet beetles and Indian meal moths.

Droppings. Their feces have pointed ends and are about 1/4 inch long; they are similar to house mouse feces.

Habits. These mice are active year round, but outdoor populations become inactive during extremely cold weather. They will leave soiled nests to build new ones in a different location. They will establish nests in furniture, cabinets, drawers, and boxes in basements and attics, and at the juncture of the foundation and floor joists, especially at the corners. Deer mice will nest in parked vehicles and chew on electrical wires.

VOLES, MEADOW MICE

Body is blackish brown above and grey below. The head and body length is five to eight inches; the tail is about two inches long.

Food. They eat plant material, including seeds and underground tubers, as well as grass and clover.

Droppings. The fecal droppings are elongated and dark colored; they are similar to mouse droppings.

Habits. They make extensive runways below the soil surface; after snow melt, their above-ground tunnels can be seen.

HAIRY-TAILED MOLE, EASTERN MOLE

The hairy-tailed mole is five to seven inches long; the eastern mole is larger and has a short, hairless tail and webbed toes. There are no visible eyes.

Food. They eat earthworms and insects found near the soil surface.

Habits. Moles spend most their time below ground. Their underground habitat limits the predators that attack them, and populations can become large. They have one litter a year with two to six young. Adults can live for four years. Hairy-tailed moles are active above ground during the day and often are caught by house cats; they also can be caught in rodent snap traps.

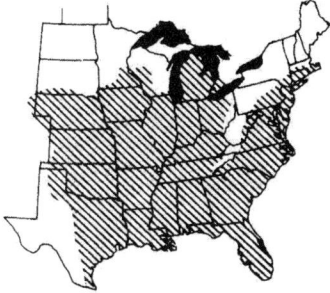

Distribution. The hairy-tailed mole is distributed in northeastern states, from the mountains of Tennessee into Canada, and in coastal regions.

The eastern mole is distributed over most of eastern U.S., from the Gulf Coast north to Michigan, but it is scarce in the northeastern states above Pennsylvania.

Eastern mole distribution

EASTERN CHIPMUNK

Body is reddish brown, belly is pale white; one white stripe on the sides is bordered by two black stripes. Ears are prominent. Length is about 10 inches, weight is about five ounces.

Food. They eat primarily of grains, nuts, berries, seeds, mushrooms, and insects. They cache food in their burrows throughout the year.

Habits. Chipmunks are primarily a nuisance problem; they can cause structural damage by burrowing under patios, stairs, retention walls, or foundations. They also may consume flower bulbs, seeds, or seedlings, as well as bird seed, grass seed, and pet food. They often are infected with Lyme disease and are a host for the larvae of deer ticks.

Distribution. The range of the Eastern chipmunk is from southeastern Canada and northeastern U.S. west to Oklahoma.

Hairy-tailed mole

Hairy-tailed mole distribution

Chipmunk

GROUND SQUIRREL

Body is brown with 13 alternating brown and pale white stripes; length is six to 12 inches and weight is about 10 ounces. Each front foot has four toes with long, digging nails.

Food. About 50% of their diet is grasshoppers, wireworms, caterpillars, beetles, ants, and earthworms. The vegetative portion of their diet includes seeds, green shoots, flower heads, roots, vegetables, fruits, and cereal grains.

Habits. They dig up newly planted seeds, clip emerging plant shoots, and feed on garden vegetables. These ground squirrels invade golf courses, lawns, athletic fields, and similar open grassy sites to burrow and feed. The openings to their burrows often are hidden, and there may be no surface mounds. Burrows are 15 to 20 feet long with several side passages and about two feet below ground.

Distribution. This ground squirrel occurs throughout much of central North America, from Canada south to New Mexico and Texas.

TREE SQUIRRELS

Tree squirrels occasionally use house and building attics as nesting sites or for food storage, or they simply enter attics as a part of their foraging behavior. Once inside, they may move into spaces between wall voids and floors. Squirrels gnaw entryways along edges of attic louvers, or gain access through vents or construction gaps under eaves and gables. There may be openings around the junction of the roof and chimney they can use to enter. The evidence of squirrel activity in an attic includes feces, nest materials such as leaves and branches, and chewed nuts and pits scattered in the area close to the entry. An inspection for squirrels requires careful examination of the roof line for entry points. Observations should be done in the morning and afternoon; squirrels are often inactive during the middle of the day.

GRAY SQUIRREL

Adult body is gray above, with a pale gray underside. Tail is gray with silvery-tipped hairs. A black phase is common in northern parts of the range, and there are albino populations.

Food is primarily nuts, especially acorns and walnuts, but they also will eat the seeds of maple and tulip trees. Nuts are buried individually and not in a cache. The buried nuts are sought in winter through detection of their scent, because they are not buried very deep.

Droppings are about 3/4 inch long, dark brown to black and slightly spindle shaped with one end pointed. There may be pieces of nuts or seeds in the feces.

Habits. Adults are active early and late in the day. Mating is in midwinter, and a litter of two or three young is born in spring, with a second litter in late summer. This species does not hibernate; it is active in the winter. Young are born in the nest, weaned in about seven weeks, and leave the nest to forage on their own in about 10 weeks.

Ground squirrel

Ground squirrel distribution

Squirrel

Squirrel feces

BATS

Bats usually roost in buildings that are near streams, lakes, and ponds; these sites will have populations of gnats, mosquitoes, mayflies, and moths on which the bats feed. Roosting site selection may be based on location for food and high temperatures suitable for rearing young. After dusk, bats start leaving the roost, and most will be out within an hour. They then move to feeding and drinking sites. Females caring for young may be gone only a few hours, but males are usually out all night. Males that do not return to the roost site use a "day roost," which may be a carport, porch, or behind shutters of a house.

Reproduction. Bats leave the overwintering site and enter structures in spring. This is usually April in northern regions, but earlier in the South. These bats are females preparing to give birth. Baby bats are born during June and July. Little brown bats usually have one pup per female, but big brown bats have two. Young are breast-fed for about seven weeks. Bats remain in their original roost all summer. Mating occurs in fall, but actual fertilization does not occur until the following spring. In fall, when temperatures decline, bats leave their summer roosts and travel to their overwintering sites. Brown bats generally live four to 10 years.

Bat in flight

Inspections. The best time to inspect for bat infestations is at dusk, when the bats are emerging. Bats emerge each evening, unless the weather is extremely adverse. Begin the inspection just before dusk and continue for about one hour after the last bat emerges. The most common exit and entry points are attic louvers, and openings of three-eighths inch or larger. Exit and entry points may be marked by smudge marks or bat droppings below the site. Indoor inspections begin with the area around the entry and exit points; bats may roost on rafters and in wall voids and may not be immediately visible.

Exclusion. Measures planned for bat exclusion must follow state regulations. In many states, bats are protected and may not be killed unless rabies is suspected. Live removal of bats through exclusion is usually the only allowed method. Some states require that exclusion and sealing of entrance holes be done between September and February; bats cannot be excluded during summer because there may be flightless young bats present.

Personal safety. Bats are responsible for numerous cases of rabies every year in the U.S. Precautions should be taken whenever work with bats is planned. A physician can provide a pre-exposure vaccination against rabies for technicians that regularly work with bats. The rabies virus is transmitted through bat bites and scratches. Always wear heavy-duty gloves when working with bats. Bat roosts that have an accumulation of droppings also can present a problem. The fungal spores that cause histoplasmosis can be present in the dried feces. A respirator is recommended if feces are moved.

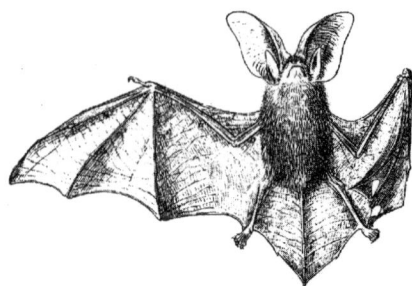
Bat

Bat feces

Big brown bat distribution

Raccoon

Raccoon footprints

LITTLE BROWN BAT

Adult body is uniformly dark brown and slightly glossy, and light grey underneath. Body length is about four inches, and the wing span is about nine inches.

Food. These small bats eat moths, midges, mosquitoes, and mayflies in flight. However, there is limited potential for bats to provide effective control of mosquitoes.

Droppings. Their feces are about 1/2 inch long, black, and slightly glossy. The ends are blunt, and insect fragments usually are visible on the outside. The feces pellets crumble easily when pressed.

Habits. Adult males and females live separately, but come together in fall then migrate south in the winter for mating and hibernation. Young are raised in nursery colonies in secluded locations, such as attics of heated buildings. Females have one baby per year, born in May to July. Young learn to fly within three weeks; by about four weeks, they are adult size.

Little brown bat distribution

BIG BROWN BAT

Adult body is uniformly dark brown to copper without distinct markings. Body length is about five inches, and the wingspan is about 14 inches.

Food. These bats eat beetles, wasps, flies, moths, and other relatively large insects. They do not feed in winter but depend on stored fat reserves.

Habits. This species is commonly encountered because of its year-round use of buildings. They are hardy and capable of surviving sub-freezing temperatures. They frequently remain active until November and may relocate in summer if their roost temperatures exceed 95° F. Their flying speed is recorded at 40 mph, the fastest for any bat species.

RACCOON

Adult is blackish brown above, with a gray underside. The bushy tail has four to six alternating black and gray rings; the head has a black mask outlined in white. It is about two feet long and weighs 12 to 48 pounds.

Food includes fruit and vegetables, also insect grubs, earthworms, crayfish, frogs, fish, and bird eggs; around houses and commercial buildings, they will eat cat food, garden vegetables, and garbage.

Reproduction. Mating is in January to March. A litter of three to five young are born in April or May. The young usually disperse in fall.

Habits. Raccoons are primarily nocturnal but can be active during the day. Most daily movements are within a relatively small area. Male home range is two to three square miles. Raccoons have short life spans; 50% to 70% of all populations consist of individuals younger than one year of age.

STRIPED SKUNK

Adult is black with two broad white stripes on back, extending to the head. Color can vary from mostly black to mostly white. It is about two feet long and weigh six to 14 pounds.

Food. Adults and young eat insects, especially grubs in turfgrass, as well as earthworms, snails, plant material, carrion, and garbage.

Reproduction. Mating takes place during late February and early March, and young are born in late April and May. The young usually disperse during the fall of their first year; males are usually solitary.

Habits. Skunks are primarily nocturnal, but can be active during the day. They may dig their own burrows, but they prefer to use natural cavities among rocks, or under stone walls, logs, or buildings. Skunks produce a strong-smelling liquid from scent glands. These glands are located on either side of the rectum. These glands secrete a sticky, yellow fluid; the main component of which is butyl mercaptan.

PIGEON

The body color of this wild bird is gray. It has a white rump and rounded tail, usually with a dark tip. The wings have two black bars. The sexes look alike; the male is larger with a more iridescent neck coloring. Size is about 14 inches.

Food. They are seed eaters; they feed by swallowing seeds which then are stored in the crop and later crushed in the gizzard. They feed quickly, and then fly off to digest the food.

Reproduction. Pigeons generally nest singly on flat areas such as building ledges, air conditioning units, or window sills. Females will lay one or two eggs which hatch after approximately 18 days. Normal life span is about four years.

Habits. Pigeon ectoparasites, such as mites, may invade homes from pigeon nests in or on the building. *Salmonella* is found in about 2% of pigeon feces and can cause *Salmonella* food poisoning in man. Pigeon droppings deface and accelerate deterioration of buildings. Histoplasmosis and cryptococosis are systematic fungus diseases in humans which can be contracted from dusty pigeon manure.

Skunk

Skunk footprints

Pigeon

Woodpecker

WOODPECKER

Body can be colored red and black, and the wings can be banded. The common species are eight to 12 inches long. The beak is generally long and sharply pointed.

Food. They are primarily insect eaters, and search for insects and larvae that may be below the surface of wood. These birds can detect insect activity, such as carpenter bees and larvae, in galleries in wood siding or other exposed pieces of trim.

Reproduction. Breeding for all woodpecker species is in spring; females lay three to six eggs, which hatch in about 11 days. Both the male and female tend the young; there may be several broods each year.

Habits. Siding, such as cedar, is attractive to woodpeckers. When a woodpecker is looking for food, it will leave several half-inch diameter feeding holes. These holes may be formed into rows, and are often seen near the eaves of the house. One or two larger holes, one inch in diameter, are usually a sign of nesting.

SPECIES INDEX

C

D

E

F